D0773585

THREE MILE ISLAND

THREE MILE ISLAND

Mark Stephens

Random House   New York

Copyright© 1980 by Mark Stephens

All rights reserved under International and Pan-American Copyright
Conventions.
Published in the United States by Random House, Inc., New York, and
simultaneously in Canada by Random House of Canada Limited, Toronto.

Library of Congress Cataloging in Publication Data

Stephens, Mark.
Three Mile Island.

1. Three Mile Island Nuclear Power Plant, Pa.
2. Atomic power-plants—Accidents. I. Title.
TK1345.H37S74   621.48′35′0974818   79-25214
ISBN 0-394-51092-5

Manufactured in the United States of America

2 4 6 8 9 7 5 3

First Edition

Foreword

David Rubin of New York University hired me to work for the President's Commission on the Accident at Three Mile Island. As a member of the Public Information Task Force, I worked with Holly Chaapel, Ann Marie Cunningham, Mary Beth Franklin, Sharon Friedman, Wilma Hill, Nancy Joyce, Roy Popkin, Peter Sandman, Patricia Weil and Emily Wells. Their probing, questioning and insight made this a story. During those ninety-hour weeks, Emily Wells was my sanity.

Vital material was provided by Chuck Harvey, Ruth Dicker and Dan Reicher of the Commission's Office of Chief Counsel. Their efforts give the book much of its texture and, I hope, balance.

Back in California, a small army of students helped me plow through the thousands of pages of documents and notes that were the raw material of the book. Sally Marone Baird, Achla Bedi, Laurie Bennett, Brad Glover, Joy Murakami, Patricia Rowell and Rosemary Storm were my eyes and, in some cases, my brain. Robert Briller, an electrical engineer turned doctor, led me through the technical maze of the accident and acted as interface with many experts at the Electric Power Research Institute.

Dr. Roland Finston of Stanford's Department of Health Physics and Dr. Rudolph Sher of the Department of Mechanical Engineering often set me straight and pointed me to just the right experts.

Hundreds of reporters, editors, producers, engineers, politicians, bureaucrats, scientists and citizens answered my questions with varying degrees of belligerence, fear, cooperativeness, eagerness, anxiety and disinterest.

The IBM 3033 computer upon which the book was written *almost* never failed.

Finally, Robert Loomis, my patient editor at Random House, provided just the right levels of insistence and understanding to make this writer, just maybe, into an author.

M.S.
Palo Alto, California

Contents

CONTENTS

THREE
MILE
ISLAND

Prologue

Everyone noticed the tension. The CBS Broadcast Center was even busier than usual as the 5:30 P.M. taping time for the CBS Evening News with Walter Cronkite approached. Where normally the news program is set and ready to go a half-hour before taping time, here the technicians were standing ready, waiting for any new piece of tape or film that might drop in to their West 57th Street location.

Cronkite was in his glass-walled office, just off the small news studio. Books lined two walls of the cubicle, jammed together on the shelves. Books on space flight, on sailing, on many favorite subjects. Just behind the desk glistened an Emmy award, in sharp contrast with the dog-eared books.

It was time to comb Cronkite's silver hair and time for glasses to come off and contacts to go in. A dozen steps and into the chair where millions of American viewers would see him tonight. The lights came up, blinding. Camera positions checked, colors stabilized, a dab of powder on a shining nose. Writers dropped into the desks on either side of the Old Man, surrounding the managing editor of the CBS Evening News. They sat in shirt-sleeves as background to the aging man in the blue Brooks Brothers suit. Their typewriters were IBM Correcting Selectric II's and unused. In the real newsroom, the typewriters are manual.

"Ten seconds!" A final shuffle of paper and the program was ready to go. There would be no last-minute piece of film from Pennsylvania after all. A swell of taped teletype sound rose in the background,

3

mimicking the actual printers that churned out paper in the next room. The scene was Cronkite alone as the intro ended. America was watching the man most trusted of all Americans. It was March 30, 1979.

"Good evening. The world has never known a day quite like today. It faced the considerable uncertainties and dangers of the worst nuclear power plant accident of the atomic age. And the horror tonight is that it could get much worse. It is not an atomic explosion that is feared; the experts say that is impossible. But the specter was raised that perhaps the next most serious kind of nuclear catastrophe, a massive release of radioactivity [could occur]. The Nuclear Regulator—Regulatory Commission cited that possibility with an announcement that, while it is not likely, the potential is there for the ultimate risk of a meltdown at the Three Mile Island Atomic Power Plant outside Harrisburg, Pennsylvania. Robert Schakne reports from Washington. . . ."

It was fitting for a day such as "the world has never known" to be reported in this way. For the first time in the history of the Columbia Broadcasting System, the CBS Evening News devoted nineteen minutes to a story. Up to that time, no single news story had ever received such intense coverage—no war, assassination, disaster, election, trial or hearing. That night, Three Mile Island beat Dallas, 1963, and Saigon, 1975. The economy was ignored, the Boat People forgotten.

What seemed to reach a peak of intensity, uncertainty and fear that Friday night had begun two days earlier 150 miles from the Manhattan studios in a place few Americans had ever seen or even heard of. That earlier day, March 28, 1979, saw the beginning of the worst nuclear accident in history, the worst threat to public health of the nuclear age.

Following a dozen less serious accidents, a hundred near-misses and a thousand malfunctions, the nuclear industry had that day to face up to conditions it had claimed were "impossible." And the people of Pennsylvania and of the world found themselves having to stand by, accepting the impossible and hoping that the men who had been so wrong in claiming that a serious accident could not happen

4

would be able to deal successfully with the accident now that it had come.

The story of Three Mile Island is generally one of information—its presence, absence and evasive nature. Technicians, engineers and, later, politicians, were all seeking information about what was happening inside the reactor core of Three Mile Island's Unit 2. For a long time they were unsuccessful in getting this information. For part of that time they pretended it didn't matter.

While engineers were searching for the truth, so was the press. Facing a new type of event to cover on an otherwise-slow news day, the press tried to understand the inner workings of the reactor as well as what was going on in the minds of those trying to deal with the problem and of those who had to live in the shadow of Three Mile Island—a plant that didn't even generate electricity for them to use.

It is the story of twenty-six agencies and their heads, all trying to come to the rescue of Three Mile Island, or at least to appear to be doing so. It's a story of confusion, waste and ineptitude in a search, often not for information, truth or success, but for public attention.

The press was at the center of Three Mile Island. They were charged with interpreting the event for the nation. Strangely, for accepting this task, they came under the most criticism of any group involved—criticism for sensationalism, for being antinuclear, for being pronuclear and for not keeping reactor specialists on the reporting staff. While the press was searching for information, some involved agencies tried to provide it and others tried to hide it, even to the point of lying to the press and, through them, to the American people.

This book is an attempt to show what really happened during the accident at Three Mile Island. It covers what was taking place in the reactor itself, how utility and public officials responded and how the press tried to do its job. It is an attempt to show the inner workings, and sometimes failings, of government trying to react to the crisis.

Some who read this book will find it to be antinuclear. Others will swear it is pronuclear. It is meant to be neither. Rather than supporting or condemning an energy source, the book looks at people, agencies, institutions and ideologies interacting against the backdrop of a particular technology. And while no recommendations about the future of that technology are made here, a few conclusions may be found about the way men ought to manage it on a day-to-day basis.

During the accident at Three Mile Island, a hundred thousand people were almost exposed to excessive doses of radiation, because men in power within both the private and public sectors, through fear, greed or incompetence, put politics, economics or pride before the public health and safety. Faulty pipes, valves and instruments can be repaired or replaced, but is there a repair for reckless self-interest? Mistakes were made on *all* sides during the accident at Three Mile Island, and many of those mistakes could have been avoided.

Only luck saved the people around Three Mile Island. But we can't continue to rely on luck.

1

A Blast of Steam

A blast of superheated steam, 500 degrees hot and under pressure of a thousand pounds per square inch, shot from a safety valve at the top of Three Mile Island's Unit 2 reactor building. A few workers on the graveyard shift at TMI heard the sound, and so did Holly Garnish, who lives a quarter mile from the plant, just up the bank of the Susquehanna River. "Picture the biggest jet at an airport and the sound it makes," she recalled. "That's what I heard. It shook the windows, the whole house."

Holly's husband John continued to sleep. They had heard the same noise more than forty times in the year since Unit 2 went critical—since the reactor had started to consume its nuclear fuel. Holly looked at the alarm clock next to her bed. It said 4 A.M. "I remember because I put the dog outside," she said.

Holly Garnish and a few others sleeping in country houses along the Susquehanna were among the first to know about the nuclear accident at Three Mile Island. Only they didn't know. They thought it was just like any of the other blasts of steam that had shaken them before. And Holly wasn't the only person that night to think little of the steam. It didn't cause much concern at first in the Unit 2 control room, either.

The turbine building had been very warm at 3 A.M.—nearly 100 degrees Fahrenheit. And there was so much humidity in the warm air that the walls dripped with condensation. But these conditions

7

were not unusual. Nuclear power plants are places of heat and water and, especially, of sound, with the steamy smell of laundry day altered by the incredible noise of millions of gallons of water pumped through pipes three feet in diameter and suspended just overhead. Every now and then, a crashing bang would ring out, a sound that anywhere else would mean a serious problem, but in the concrete containment building was just another normal sound of water and pipes. The three workers didn't even look up.

In and around the massive plumbing of the heat exchangers and the steam generators of the secondary cooling system loop, shift foreman Fred Scheimann and two auxiliary operators were working on condensate polisher number 7. Really a set of filters to make sure that only clean water without dissolved minerals would be made into steam for the turbines, the condensate polishers were giant columns filled with special resins.

Filters get dirty, and it was time for the resin columns to be flushed and renewed. That is what this crew of men and the crew before them had been trying to do for eleven hours. They were trying to pump the resin from polisher number 7 into the flushing tanks and then back. But the resin didn't want to go.

Tired and exasperated, the men swore at the blocked pipe connecting the polisher with the flushing tank. There was a plug of resin stuck in the pipe and nothing would move it. They had been using compressed air to push the resin and found that the 100 pounds of air pressure available to them just wasn't enough to move the honey-thick resin.

Changing strategies, Scheimann used air to agitate the water in the bottom of the polisher, hoping that the water would help loosen the lodged resin. Just as a child blows through his straw to make bubbles in a glass of milk, the operators blew bubbles in the water. The resin column was like a milk shake, bubbled full of air and water and resin. Finally the resin blockage was pushed free.

Scheimann's crew went on to other work around the plant. Behind them, though, they left something—a small amount of water— forced past a leaking valve on the side of the condensate polisher by the pressure of the churning water.

8

The water drifted through the instrument air system of the plant. This is a complex of miles of pipe running throughout the giant plant and powering pneumatic controls on a wide variety of machines. Eventually some of the water was blown into control valves on the feedwater pumps to the main turbines. These valves, controlled by air pressure, automatically close if air pressure is lost or if the flow of air is interrupted in any way—interrupted by water, for instance. Less than a glassful of water was about to start the accident at Three Mile Island.

Just one minute before Holly Garnish would wake up, three men sat in the control room of Unit 2. It was 3:59 A.M. They were matched by other men in Unit 1, the other reactor on Three Mile Island. But Unit 1 was down for refueling and its operators had little to keep them busy. Craig Faust and Ed Fredrick were in the control room, with Bill Zewe in his little office just a few feet away. Zewe was the shift supervisor. The shift foreman, Fred Scheimann, had not yet returned from working on condensate polisher number 7.

Faust and Fredrick sat at their desks in the blue-carpeted control room, a windowless forty feet square, dominated by the big horseshoe-shaped control panel. Really a double panel, the bank of instruments, lights and controls had an inner console with a taller outer console running parallel behind it. Hundreds of lights and vertical gauges covered both units. And many of the instruments were marked with sticky plastic labels or with masking tape and ball-point pen. Some controls carried paper tags put there as a warning during a maintenance procedure or as a safety precaution.

Not all the plant functions were indicated on the big U-shaped panels. Some functions were monitored by a computer and printed out at a typewriter near the control board. And some of the controls were on a back panel behind the main consoles and out of sight of the operators. These out-of-sight controls would become important in a matter of seconds.

Bill Zewe was in his little glass cubicle of an office off the main control room. Through the windows he could see the other men. Like Zewe, they were all products of the nuclear Navy and all young

—the oldest no more than thirty-five. Youth and Navy experience are common characteristics of commercial reactor operators in the United States.

The control room was quiet with the plant running normally at 97 percent power, generating enough electricity to light 300,000 homes; the Integrated Control System, a computer to run Unit 2, was on full automatic; it was 37 seconds after 4 A.M.

Two feedwater pumps failed in the secondary cooling system of Unit 2. There was no more water circulating to remove heat from the reactor. The temperature and pressure of water within the reactor soared. An Electromatic relief valve on the pressurizer tank opened, shooting steam into a storage tank in an attempt to lower the pressure. At the same time, the turbine "tripped" or stopped and another valve released steam into the outside air—steam that would have been used to generate electricity but, instead, shrieked into the night to be heard by Holly Garnish.

Within nine seconds, 69 boron and silver control rods fell into place among the 36,816 zirconium fuel rods with their millions of pellets of uranium dioxide fuel. The rods absorbed neutrons to stop the chain reaction. The falling into place of these rods, called a "scram," worked as it should. The reaction stopped. And the Electromatic relief valve on the pressurizer reduced pressure in the cooling system to a safe level. Then the relief valve should have closed again and shut off the flow of steam. But it didn't.

The valve stayed open. It showered radioactive coolant into holding tanks. When the tanks were full, a safety plate burst and allowed the steaming water to flow onto the floor of the fortresslike concrete containment building. Pumps kicked on automatically to transfer the coolant to holding tanks in the auxiliary building next to the reactor. The control room operators, already busy with other things, did not know that the Electromatic valve had not closed. Their instruments said it had closed, but only measured the performance of a faultily closing solenoid, rather than the position of the valve itself.

Electromatic valves are expected to fail once in every fifty times they are activated. This particular valve had failed once in 1978 and was leaking even before the accident—allowing six gallons of cooling water to leave the system every minute. Nuclear Regulatory Commission Chairman Joseph Hendrie said later: "These Electromatic valves are designed to open every time and under any conditions, but they're not designed to close under any conditions." That's why there is a motor-driven blocking valve, just like a faucet, also mounted on the pressurizer to shut off the flow from a failed Electromatic valve. The blocking valve remained open.

With the Electromatic valve stuck open, the pressure in the cooling system continued to drop. And the system temperature continued to rise. Even though the reactor had scrammed, the reactor was still producing plenty of heat. Radioactive products of the nuclear reaction were decaying, converting spontaneously to different forms, with an accompanying release of heat as they slowly lost their radioactivity. With the reactor completely shut down, this decay heat was still producing at least 6 percent of normal reactor power —enough to light 18,000 homes.

At the lower system pressure, the super-hot cooling water started to turn to steam. Bubbles of steam began to form in the cooling loops. The steam generators, like teakettles left too long over the flame, began to boil dry. They needed a new source of cool water within 90 seconds to avoid damage.

Within a few seconds, three auxiliary feedwater pumps kicked on automatically to reestablish feedwater flow to the steam generators. The accident was still only 14 seconds old. Yard-thick feedwater pipes resounded and swayed under some tremendous force. Water hammer! Just as home plumbing can groan and bang when a tap is quickly closed, the mammoth plumbing of Unit 2 banged with a force that is known to destroy valves and crack brittle piping. Closed valves were blocking the flow of water from the just-started auxiliary feedwater pumps, so the pumps were now beating their brains out, trying to force water into the closed-off system. And the steam generators continued to boil.

The two valves that prevented the auxiliary feedwater pumps from doing their job were not supposed to be closed. In fact, operating with these valves—called the "twelves" by reactor operators—closed is a violation of rules set by the Nuclear Regulatory Commission. The valves had been shut for two days since a test of the auxiliary feedwater pumps on Monday. A few days later, one operator would say, "It's easy to forget to open the valves. You can get distracted or you may be relieved in a change of shift."

Back in the control room, with the accident still less than a minute old, the operators were very busy but also confident that they had the emergency in hand. At the original valve failure, an alarm rang out in the control room. Faust and Fredrick looked to the alarm board, a large indicating panel mounted above the control stations with lighted squares called "annunciators" for each of the 1,600 failure points monitored in the system. Though the alarm board had 1,600 lights, there were only three audible alarms in the control room —one for the alarm board, another for the computer and a third for the radiation monitoring system. Turning off the audible alarm might turn off some of the 1,600 indicator lights, so Faust decided to let it ring until the situation was under control. He did not expect that the alarm would ring all morning.

The design of the alarm board was a weakness in Unit 2. The annunciator "windows"—lighted panels with a written description of the failure—were difficult to read. And if the operators cut off the sounding alarm and the particular condition that had lighted one or more of the alarm windows was no longer present, the light behind that window would go out, leaving no record of what sort of failure had taken place. That was why Faust left the audio alarm on: so that the history of the accident would not be lost.

Ed Fredrick especially disliked the alarm board. In May 1978 he'd written a memo to Metropolitan Edison, the reactor's owner, pointing out problems with the board and suggesting possible solutions. No action was taken or planned on Fredrick's memo. After the accident he said, "There were times I wished I could pull that alarm board right off the wall."

At first the operators thought the problem was just a turbine trip —a malfunction of the giant generators—and they began the emergency procedures specified for that occurrence. But then shift supervisor Zewe came out of his office as another light appeared on the alarm board. "You just lost the reactor," he said to Faust.

That initiated a different emergency procedure. "We moved from turbine trip to reactor trip," Faust said later. "The verifications there are different. So I verified that all the rods had dropped into the core. The individual rod positions are straight ahead on the panel, and so I verified that and the neutron power indication coming down."

In a graphic display, the rod status panel showed lines of red lights to symbolize the 69 control rods. As the reactor was shut down, these lines of lights began to turn on, following, from top to bottom, the descent of the control rods into the reactor core.

Faust then moved on to the secondary cooling system where he verified that the main feedwater pumps had tripped and the auxiliary pumps had turned on—the pumps that were kept from doing their job by the closed "twelve" valves. While Faust verified that the auxiliary feedwater pumps had started, he did not see warning lights that indicated the two valves had closed off the system. One of these indicator lights was covered by a paper warning tag. It is still not known why the other light wasn't seen.

The general plan of action was to stabilize the plant, keeping the coolant temperature at about 547 degrees and the system pressure at 2,155 pounds. With the system in this state, the decay heat would be easily absorbed and removed from the reactor.

But the coolant temperature wasn't 547 degrees. It was 575 degrees and rising fast. And the system pressure was dropping rapidly. By this time, the water level indicators for the steam generators showed that they had boiled dry! Things were definitely not following the book.

When the accident began, it was Ed Fredrick's job to maintain the water level in the system pressurizer, a tank of water and steam used to maintain pressure and water level in the cooling system. When pressure gets too low, electric heaters in the pressurizer heat the water there to increase the size of the steam bubble that maintains system pressure. And when pressure gets too high, water sprays cool down the pressurizer and reduce the size of the steam bubble. When the reactor scrammed, water in the cooling system should have lost heat very quickly—so quickly, in fact, that the cooled, denser water

13

would have contracted and drained completely out of the pressurizer, uncovering the electric heaters that would automatically turn themselves off to prevent damage. This is not what Fredrick and Faust wanted. They needed the pressurizer to control the shutdown of the reactor cooling system, so it had been important for the makeup pumps to cut in at the beginning of the accident, adding fresh water to make up for the expected lowering of pressurizer level.

But now they had just the opposite problem. Where they had expected the pressurizer level to drop, it was rising. There was too much water in the pressurizer. "It reached about 385 inches, 400 being the limit," Fredrick said. "It was my concern not to let the pressurizer fill solid with water, because a solid water system is difficult to control and we had never done it before."

Fredrick took the makeup pumps, the high-pressure injection pumps of the emergency core-cooling system, out of automatic by pressing 6 bypass buttons on the console to his left. He closed 2 of the 4 injection valves and shut down one of the 2 pumps that were filling the system, cutting the flow of water in half. "I saw a slight tapering [of the level increase]," he said, "and then it went right back on the same rate." Fredrick called out to Zewe, "We're about to go solid!"

The operators thought at this time that there wasn't a leak in the cooling system at all and that their emergency actions, in turning on the makeup pumps, had overfilled the pressurizer. They reacted to the new situation by cutting the flow of water into the system and opening a valve to drain some water out through the condensate polishers. This is not what they should have done.

While Fredrick was cutting back the flow of makeup water and increasing the system drainage, the cooling system was rapidly heating up. Since no heat was being removed by the steam generators, the system was very hot and at a low pressure, too. The open Electromatic valve made it impossible to maintain system pressure and that, combined with the high temperatures, had allowed large bubbles of steam to develop in the cooling loops, effectively ending all circulation of coolant and forcing what water was left in the system out through the only path available—up through the pressurizer and

14

out the Electromatic valve. It was like a heart pumping away its own supply of blood. And Fredrick had just cut off the transfusion.

Out came the books. Emergency Procedures, EPs in the language of the operators, were scattered over the desks that filled the inside of the U-shaped console area. Zewe, Faust, Fredrick and now Scheimann scanned the forty feet of control board, looking for some instrument or control that was obviously out of place. Emergency Procedures dictate the professional lives and actions of nuclear reactor operators. Written by engineers, the EPs are expected to give the answer to any potential problem with the reactor. Not only don't the operators have to think for themselves in using the EPs to respond to an accident situation, they are encouraged not to think at all, believing that there is an EP for every type of accident. There was no appropriate EP for the accident unfolding at Three Mile Island. This made the operators rely on their own background, training and intuition. So pervasive is the effect of programmed emergency procedures in the nuclear industry, the operators are not trained to actually operate their reactor, but to pass oral and written tests based on the Emergency Procedures. There is no significant attempt to teach understanding of the system and the ability to reason through accident problems. Rather, the operators, who are required only to have a high school education and no engineering or physics training, are schooled in how to take a seven-hour written and a seven-hour oral examination, responding with rote answers from the Emergency Procedures. Go beyond the EPs, and most reactor operators are lost.

The high-pressure injection pumps had been pouring a thousand gallons of water per minute into the reactor core until Fredrick had shut that flow in half. With the amount of decay heat being produced, it was later calculated that at least 400 gallons per minute were required to prevent damage to the reactor core. Despite the actions of the operators, Unit 2 still had a chance of survival. If the stuck Electromatic valve had been discovered and closed at this point or if the high-pressure injection had continued to provide at least 400 gallons per minute to the reactor core, TMI Unit 2 might have gone back into the business of generating electricity in a few days.

Still only three minutes old, the accident had already destroyed one of the two steam generators. System pressure dropped to a low of 1,350 pounds, coolant continued to escape through the Electromatic valve on the pressurizer and the auxiliary feedwater pumps were pumping their needed water nowhere at all. What Faust and Fredrick thought was a cooling system filled with water was really just a cooling system filled with steam. Still trying to control the water level in the pressurizer, Fredrick turned off the second of the two high-pressure injection pumps. Water stopped flowing to the reactor core and Unit 2 was doomed.

"The temperature was high," Fredrick said later. "We had high temperature and low pressure, and we had a full pressurizer. These things were conflicting. We were talking about it, trying to figure out what was wrong. We began to distrust the pressurizer level instrument."

Whether they trusted the pressurizer level instrument or not, the operators continued to follow procedures as though it were reading correctly, denying their suspicions and continuing to hold back the high-pressure injection water that the reactor so desperately needed while still allowing the system to drain.

Eight minutes into the accident, Craig Faust was scanning the lights and instruments of the secondary cooling system. This time he saw one of the lights that indicated the auxiliary feedwater cutoff valves—the "twelve" valves—were closed and had been shutting off water to the steam generators.

"The twelves are closed!" Faust cried out, causing the other three men to look up in surprise from their work. In Faust's hurry to turn the fist-sized L-shaped valve controls, "he nearly ripped them out of the panel," Fredrick said later.

Feedwater was finally reestablished to the secondary cooling system, though either overheating or thermal shock had opened a dangerous leak in one of the two units, leaving just a single steam generator to remove heat from the reactor.

More than a hundred alarm lights glowed on the alarm panel. The single audible alarm continued to sound. The computer was typing out warnings, problems, alarms of its own. The integrated control system monitored many functions in the power plant and had already noticed the high tailpipe temperature that might have indicated that the Electromatic relief valve was still open. Unable to do anything about the valve, the computer sent an order to print out a warning on the typewriter terminal in the control room. But this warning was just one of thousands that were waiting their chance to be printed. It would be three hours before the computer would print out a warning about the Electromatic valve.

With the high-pressure injection pumps turned off and no water entering the reactor core, the coolant vital to safe operation of the reactor began to boil away. Just like the steam generators had before, the reactor core was now beginning to boil dry.

The process was slow. There was some cooling being provided by the main reactor coolant pumps transferring heat out through the one functioning steam generator, but the main problem was just that no additional water was being sent to the core. Normally the reactor core is covered by at least six feet of water. Very slowly, the water level over the core began to drop.

Back inside the control room, it looked like conditions in the reactor had finally stabilized. Temperature and pressure readings from inside the reactor vessel were reassuring. They were still far from normal, but had markedly improved, so the operators turned to other problems about twenty minutes into the accident.

Operators from the control room of Unit 1, Three Mile Island's other reactor, came over to help out. Their reactor was down for refueling. Zewe had one of these men call Gary Miller, head of both reactors and the station emergency director. Miller was at home asleep. Awakened by the call, he did some paper work for the next hour waiting for more news from Unit 2.

The level of water over the reactor core continued to descend. More steam was produced in the cooling system. The volume of water being pumped by the main coolant pumps began to go down as the steam bubbles in the system continued to grow. Coolant flow dropped to 60 percent, then 40 percent. The main coolant pumps began making noise and dangerously vibrating. The emergency procedures indicated that the temperature for operating these pumps had been exceeded. They were "cavitating"—pumping more steam than water—and tremendous stresses were placed on the 9,000-horsepower pumps. At 5:22 A.M. Ed Fredrick shut down the first pair of pumps because "we could have damaged the pump seals and developed a leak where the steam would come out through the pumps."

The first set of pumps was cut off to see if the system could continue normally without them, a test that proved to be academic. At 5:40 A.M., the other pair of pumps shut themselves down automatically and could not be restarted. The shutdown disastrously slowed the flow of water around the already damaged fuel rods in the reactor core.

While the operators were trying to follow their Emergency Procedures, the sizzling-hot zirconium cladding that surrounds the glazed uranium pellets in the fuel rods began to disintegrate as heat and chemical reactions ate away at them. Primary coolant water is always a little radioactive, but the breakdown of the cladding rapidly increased the level of radioactivity in the water as the uranium fuel was exposed and radioactive products of the nuclear reaction began to leak into the system. The two most prominent of these fission products were xenon gas and radioactive iodine that bubbled from the narrow space that had separated zirconium and uranium before the cladding had begun to disintegrate. As the water level decreased in the reactor, cladding failure increased tremendously. From being initially covered with six feet of water, the core was partially uncovered by 6:15 A.M.

The first person from outside the plant to arrive at the control room of Unit 2 was George Kunder, the station's head of technical support. Kunder lived the closest to TMI of all the top technical people. In response to an early telephone call from one of the auxiliary operators, he pedaled to the plant through the predawn darkness on his bicycle.

Kunder made a conference call to station manager Gary Miller, Jack Herbein, Metropolitan Edison's vice president for Electrical Generation, and Leland Rogers, TMI's representative from Babcock & Wilcox, the reactor's builder.

Kunder was concerned. "I felt we were experiencing a very

unusual situation, because I had never seen pressurizer level go high and peg in the high range and at the same time pressure being low," he said. "They have always performed consistently."

At 6:20 A.M., Rogers asked whether the block valve on the pressurizer—the safety shutoff valve mounted just before the Electromatic valve—was closed. Not knowing, Kunder sent one of the auxiliary operators around to the back control panel to check. This operator, whose name is not known today, came back to report, Yes, the block valve was shut. The Integrated Control System computer reported closing of the pressurizer block valve at 6:22 A.M.

The Electromatic relief valve had remained open for two hours and 22 minutes, releasing more than a quarter of a million gallons of radioactive cooling water that now filled every available space in the reactor and auxiliary buildings.

Outside, the sky was turning pink as the sun rose over the Pennsylvania countryside. And inside the Unit 2 control room, the technicians were finally realizing that something was very wrong in the 40-foot-tall stainless-steel reactor vessel.

Closing the Electromatic valve did not solve the problems in the reactor. At least eight feet of the 12-foot reactor core was uncovered, standing above the water level in a cloud of superheated steam. And the steam that surrounded the upper core provided the only cooling. Ordinarily, even this—the steam cooling—would have been enough. Computer simulations later showed that the core would not melt so long as a foot of water remained at the bottom of the reactor vessel. But this optimistic simulation supposed that there was no other damage to the reactor core. At some point, cylindrical pellets of uranium dioxide were set loose inside the steaming reactor. They tumbled in on one another, concentrating the effects of the decay heat and making core cooling even harder to achieve. At the same time, a bubble that would not be acknowledged for two days started to form above the fuel rods as water molecules split into hydrogen and oxygen under influence of the intense radiation field. The oxygen combined with the zirconium. The hydrogen rose to the top of the reactor.

With the main cooling pumps out of action, the operators hoped that the reactor would be cooled by "natural circulation" of water in the primary cooling system. The book said this should happen, but the book didn't anticipate the large steam bubbles that were keeping any water from circulating.

As the water level dropped in the reactor, temperatures in the core quickly rose. A computer printing out temperatures from the top of the core suddenly showed the temperature rising to almost 750 degrees as the uncooled reactor went crazy. There was still enough decay heat in the reactor to light more than 15,000 homes and nowhere for that heat to go. Then the computer, badly confused by a situation it was never programmed to deal with, began printing out lines of question marks where there had been numbers before. It would be hours before the utility would get around to calling Babcock & Wilcox, the reactor's builder, to ask about the question marks. They would be told it was a sign the reactor was too hot. Or too cold. Or that the thermocouples reading temperatures above the control rods were malfunctioning.

The thermocouples were part of an experimental installation in Unit 2 and one or two other commercial power reactors. Such temperature-sensing systems were not required by the NRC, so almost all commercial reactors had no measure of core temperature other than the temperature of the primary cooling water.

Sometime that morning, a technician connected a voltmeter directly to the thermocouple wiring to bypass the computer and read the core temperatures. Connecting the test instrument, the operator might have reviewed in his mind the fact that core temperatures are normally maintained at 547 degrees Fahrenheit and that the computer had last reported a very high reading of 750 degrees.

The voltmeter was attached to one of two dozen thermocouples that monitor core temperatures. Converting the reading into degrees, the operator looked up from his scrap of paper and reported a temperature of 2,400 degrees! A dozen more readings were taken and none of the thermocouples registered less than 2,000 degrees. It was decided that there had to be a malfunction in the thermocouple circuits. Actually, hot spots in the reactor core had reached a temperature of 5,000 degrees and some of the uranium fuel was melting.

It was at this point that Victor Stello, head of Inspection and Enforcement for the NRC's division of operating reactors, felt that it should have been obvious to the reactor operators that the core was uncovered and extensive fuel damage had already taken place.

"The reason for believing you had significant damage was the fact that we had a clear indication that there was superheated steam coming out of the reactor vessel," Stello said later. "The only way you can, in fact, get superheated steam out of the vessel is to have the core uncovered."

Stello felt that such high coolant temperature at low system pressure was an obvious giveaway and that "any operator who can read a steam table" should have known it.

Whether they bothered to read any steam tables in the Unit 2 control room that morning or not, Faust and Fredrick could not even imagine the core uncovered and neither did the first NRC inspectors who arrived at the plant after 10 A.M. Stello didn't share his "obvious" brainstorm.

With the core-temperature thermocouples seemingly out of action, the only reactor temperature available was from a reading of the coolant water temperature, and this temperature had finally started to fall. But the system was playing tricks again on the operators. This temperature reading, called "T-ave," was an average of coolant temperatures in the "hot leg" of the cooling system where the coolant emerged from the core and from the "cold leg" where cooling water entered the reactor. With no coolant circulating to the core, the "cold leg" temperature had dropped down to about 150 degrees, while the "hot leg" temperature had gone off the top of the scale. The average figure T-ave was calculated by the instrumentation from the artificially cool cold leg temperature and the last readable temperature from the hot leg. This type of averaging made it seem as though the system was cooling down when it was actually hotter than ever.

Up to this point in the accident, there had been no indication of a radiation problem. The radioactive steam that floated from the auxiliary building had done so undetected. But after 6 A.M., radiation alarms began to go off in various parts of the reactor and auxiliary buildings.

21

Other strange things were happening in the reactor. The concentration of boron in primary-coolant samples began to drop when it should have been rising. And the reactivity—the number of neutrons free in the core—began to go up. The borated water that was so important to stopping the chain reaction by absorbing free neutrons was being pushed out of the core by the growing steam bubbles, and the reactor was beginning to come back to life—if only a little.

The coolant sample was taken by opening a drain line in the plant's radiation laboratory—a drain line that extended hundreds of feet from the reactor itself. The drain line was so long that water samples took twenty minutes to flow to the lab. When that first sample was tested, it showed more than just a low boron count. It showed that, for at least the last twenty minutes, there had been a lot of radioactivity released in the reactor. The sample was very "hot."

Shortly after 6 A.M., a man was sent into the auxiliary building with a handheld radiation detector. He reported radiation levels of up to 1,000 millirems, many thousand times higher than background levels and an amount of radiation worthy of concern.

More radiation alarms went off around 6:35 A.M. Radiation levels in the drain line leading from the auxiliary building had risen dramatically, further showing that everything "hot" wasn't in the concrete containment dome.

The radiation alarm board in the control room was covered with amber and red lights. Amber for alert and red for alarm. Zewe said, "Just about 6:50, all of these alarms started coming in hot, just about the same time they were just going amber-red, amber-red on all indicators, and I knew that we had a tremendous problem. So that is when I made the announcement and sounded the alarm that a site emergency had been declared in Unit 2."

With a site emergency declared at 6:56 A.M., the book finally said it was time to tell the world about the situation at Unit 2.

Zewe and Kunder still did not realize that the reactor core had deteriorated to the point that radiation from the damaged fuel was "shining"—beaming through the 8-inch steel walls of the reactor vessel and the 4-foot-thick concrete walls of the containment building. But they did know they needed help. Zewe ordered his auxiliary operators to call the state Civil Defense office and NRC regional headquarters to tell them of a "slight problem at Three Mile Island."

2

One Chance in a Million

The reactor core of Unit 2 stood, glowing, in a bath of steam. The zirconium tubes that had held millions of pellets of uranium oxide fuel were twisted, split open like overcooked sausages, their contents spilled within the reactor pressure vessel.

The owners of the power plant were concerned about protecting their near-billion-dollar investment. Replacing the reactor core alone would cost $20 million, and it was obvious to many by late morning that core replacement, at the least, would be required.

But what about the public agencies, the local and state governments that were about to be contacted? And what about the people around Three Mile Island? What was the danger to them? Did damage to the reactor core necessarily mean that the public health and safety were in jeopardy?

Yes. The worst accident feared by any reactor operator is a "meltdown." Though the operators of Unit 2 would have denied it at the time, a meltdown was the worst of the possible consequences of their actions that morning. It was remote, but still possible.

Reactor core "meltdown" is caused by a loss of coolant flow as a result of a power failure or break in a major coolant line or failure of essential safety systems allowing the decay heat of the reactor to build in the reactor vessel. If there is any coolant left sitting in the pressure vessel, it soon boils away. Starting with the reactor core covered by 6 feet of water and no cooling system in operation, a disastrous sequence of events would begin which, if uncorrected, would reach their conclusion in 12 to 15 hours.

Within 45 minutes, enough water would have boiled away to expose the top of the core. At 90 minutes, the core would be half covered; at 135 minutes, it would be suspended clear of the water, cooled only by steam. Shortly after that, the zirconium cladding around the fuel pellets would begin to melt and rupture, reacting with the steam to release hydrogen gas, and the pellets would fall together, concentrating the effects of the decay heat.

The temperatures of the fuel pellets would rise until they reached the melting point of uranium dioxide, when they would begin to form a pool of molten fuel. At some point the molten uranium dioxide would melt-through—or fall through—the structure that once supported the core, falling into the last cooling water, lying in the bottom of the cylindrical pressure vessel. This would probably cause a nonnuclear explosion.

If the reactor pressure vessel with its 8-inch steel walls survived the water-uranium dioxide reaction, as it probably would, the molten fuel would form a pool in the bottom of the vessel. Eventually the puddle of liquid fuel, at a temperature of more than 5,000 degrees, would melt-through the bottom of the pressure vessel and fall into the concrete containment structure.

The molten fuel would heat the 10-foot-thick concrete floor, making it buckle and crack. Reacting chemically with the concrete, it would eat through the material and release oxygen. The oxygen would burn with hydrogen released by the water-zirconium reaction. The pressure of this burning—really an explosion—added to the high steam pressure in the closed-off containment building, would crack the 4-foot-thick containment walls and release radiation to the outside.

Once through the floor, the molten fuel would continue to burn its way through the soil beneath the plant, causing violent steam explosions and finally coming to rest, 50 to 70 feet beneath the surface, encased in a bubble of crude glass produced from the soil by the incredible heat.

25

It was once thought that the melted fuel would continue to drop into the earth, all the way through the core and out again—in China. This was called "the China syndrome."

Decay products of the fuel, released in the containment during the accident, would blow out through the gaps and cracks in the concrete walls. Nuclear engineers claim that the cracks would act to filter out nuclear particles, allowing only less-dangerous radioactive gases to escape; but there is no certainty that this would really happen, as it depends on the containment cracks being very small.

There would be no massive mushroom cloud of an atomic explosion in the event of a meltdown—little outside damage from explosions at all. But a large part of the plant's inventory of radioactive decay products would be released to the air and to the ground water. Most prominent in these decay products would be several hundred million curies of radioactive xenon, strontium, cesium and iodine, and a small amount of plutonium, picked up by the wind and spread over hundreds of square miles.

If ingested or inhaled, the strontium would concentrate in the bones of those exposed, the iodine would concentrate in the thyroid glands. In the short term, thousands would die. More thousands would die in the years following, victims of cancer caused by the radioactive fallout—bone cancer, thyroid cancer, cancer of many parts of the body. If inhaled, 130,000,000th of an ounce of plutonium 239 can cause cancer. Evenly distributed, one pound of plutonium could kill every U.S. citizen.

In less than worst-case situations, the greatest danger would be posed by radioactive iodine, accumulating in dairy products and eventually in the thyroids of consumers of those products. Health physicists in Pennsylvania were very worried that day about the presence of radioactive iodine.

WASH 1400, a 1975 Atomic Energy Commission study of the probability of reactor accidents and their results, estimated that a meltdown and breach of containment would kill 27,000 people, injure another 73,000 and cause $17 billion in property damage.

Of course, the WASH 1400 study, led by Dr. Norman Rasmussen

of MIT, also assessed the probability of such a total meltdown at one in 10 million per year. That is, with the 72 commercial power reactors operating today in the United States, there is less than one chance in 10,000 of such a serious accident happening in the next decade.

And yet, Unit 2 stood glowing in its steam in some defiance of Dr. Rasmussen's calculations. There was a flaw in those figures that had so reassuringly said we need not expect a meltdown this century—a flaw that was taken advantage of by Unit 2.

The methodology of WASH 1400 made use of "event trees"—sequences of actions that would be necessary for accidents to take place. These event trees did not assume any interrelation between events—that they might be caused by the same error in judgment or as part of the same mistaken action.

The statisticians who assigned probabilities in the writing of WASH 1400 said, for example, that there was a one in a thousand risk of one of the auxiliary feedwater control valves—the twelves—being closed. And if there is a one in a thousand chance of one valve's being closed, the chance of both valves being closed is one thousandth of that, or a million to one. But both of the twelves were closed by the same man on March 26—and one had never been closed without the other.

The odds continued to change when an operator both opened a let-down line to drain the cooling system *and* turned off the high pressure injection that was sending water to the reactor core. To Dr. Rasmussen, they were independent events, each to be assigned a probability, the product of which would be the real risk of an operator's letting the reactor run dry. But both events happened together. They were the logical outcome of an operator's thinking he was overfilling the cooling system. It was one mistake, not two, and was a thousand times more likely to occur.

The type of optimistic thinking that was the foundation of WASH 1400 pervaded the nuclear industry. So small were the chances of a major accident and so fail-safe the emergency systems, that certain accidents and situations didn't even have to be considered in emergency planning.

This attitude extended to the NRC, where it often seemed that the industry's regulators were also its greatest fans.

"Sure, every reactor that operates might have a meltdown," said NRC investigator Charles Gallina. "Every car that you drive might have a fatal accident, but you don't go and buy a car and the guy says, 'You will get 21 miles to the gallon plus your odds are one to 4,000 you'll be killed in this vehicle.' It is a true statement but you know it is information that is irrelevant at the time."

On the morning of March 28, Unit 2 shift supervisor James Floyd was taking a refresher course on the Babcock & Wilcox reactor simulator in Lynchburg, Virginia. The simulator, a model control room connected to a large computer, was intended to re-create any conceivable accident situation on command.

"When I heard about what was happening at the plant, I went right into the simulator and tried to duplicate the conditions of the accident as we knew them at that moment," Floyd said later. "I tried to duplicate the instrument readings they were getting up in Pennsylvania, but I couldn't."

Following the mindset of the nuclear industry—that certain accidents and reactor conditions will never exist—the Babcock & Wilcox computer was not programmed to simulate a solid pressurizer, filled with water as the Unit 2 pressurizer seemed to be. And with the possibility of both auxiliary feedwater valves being closed pegged at a million to one, the computer could only simulate one of the twelves failing at a time.

During licensing hearings for Unit 2 in 1977, the NRC agreed with the utility—and against those opposing the license—that it would only be a waste of time to debate the effectiveness of reactor safety systems in the face of a Type IX accident—an accident involving a major release of radiation to the environment—because such an accident could never happen. The industry was so confident that a bad accident would not take place that it felt little need to make emergency plans in case one did.

A letter of reassurance from Metropolitan Edison arrived at the Middletown Borough Hall in 1974. It read: "Even the worst possible

28

accident postulated by the A.E.C. would not require an evacuation of the borough of Middletown . . . based upon the discussion above, it can be seen that it is unnecessary to have specific evacuation routes identified for the borough of Middletown."

There were no planned evacuation routes for the borough of Middletown on March 28. Nor were there evacuation plans for any other township or city around Three Mile Island.

Backed up by the statistics of WASH 1400, Met Ed and the operators of Unit 2 were confident that theirs was not a worst-case scenario. Everything would soon be under control.

And the core of Unit 2 stood, melting, in its steam.

3

The World Finds Out

Seven A.M. and Bill Whittock was standing outside his front door with a frown on his face. Across the river was the power plant that had shaken him from sleep at four o'clock and now he stood on the bank of the Susquehanna, where there should have been only the smells of spring and of the river but there was this strange metallic taste in the air.

The taste probably meant nothing, but Bill Whittock was a retired engineer, and a cautious one at that, so he decided to give Civil Defense a call just to make sure they knew.

Back inside his home along the river, Whittock looked for the telephone number of the state Civil Defense office. The telephone directory listed "Civil Air Patrol" and "Civil Service Employees," but nothing in between. There was no telephone number for Civil Defense. So Whittock put aside the idea of making a call and went for a walk instead.

It was a cool morning of early spring with not much in the way of either wind or cloud. Things seemed pretty much normal except for the sense of metal in the air and the lack of steam coming from the big cooling towers across the river.

The Pennsylvania Emergency Management Agency—what used to be called Civil Defense—had its operations center in a reinforced concrete bunker buried under the State Transportation Department Building in Harrisburg. There, a hundred yards from the state capi-

30

tol and deep underground, a small staff waits for the next disaster.

Clarence Deller was the PEMA duty officer when the first call of March 28 came in. It was from Three Mile Island. Deller logged the call: "Call from shift supervisor at Three Mile Island stating they had an emergency in reactor No. 2 which has been shut down. There is a high level of radiation in the reactor room but no off-site release. He requested that the Bureau of Radiation Protection be notified and return the call."

Deller put in a call to the state Bureau of Radiological Protection, but there was no answer. Then he set to work on the PEMA teletype network, informing Civil Defense directors and police and fire departments of the four counties surrounding Three Mile Island.

A few minutes later Deller called the Bureau of Radiological Protection duty officer at his home. William Dornsife was the only nuclear engineer employed by Pennsylvania. Before starting work with the state in 1976, Dornsife had helped design safety systems for the Unit 2 reactor as a engineer for Burns & Roe, the power plant design firm. He knew the plant well.

At 7:05 A.M. Dornsife was standing barefoot on the cold wooden floor of his Harrisburg bedroom. "I rolled out of bed when the phone rang and in my early morning mind, I was hoping against hope that this was a drill," he remembered. "I called the shift supervisor at the plant, he filled me in on the details, then there was an announcement in the background to evacuate the fuel handling auxiliary building. It didn't hit me until I heard that. And I said to myself, 'This is the biggy.' I realized that this was really a very serious accident."

Dornsife went to work immediately, calling out his radiation monitoring team and its leader, Margaret Reilly, from their homes. Dornsife was in a hurry. "They were claiming radiation levels of eight rems in the containment," he said, "and I knew for the levels to be that high, some of the fuel had to have melted."

It would be two days before Metropolitan Edison would publicly admit that some fuel had melted.

Not much was going on in Dauphin County Control when the PEMA teletype message came in at 7:03 A.M. The dispatchers had

31

sent out an ambulance to pick up a heart-attack victim early that morning, but nothing more. John Brabits, assistant director of the center, took the message and knew that something important had happened. "Not only was it unusual for us to hear from the plant," he said, "this was only the third time they'd contacted us in five years."

Kevin Molloy was hungry. The director of the Dauphin County Office of Emergency Preparedness banged around his kitchen in the predawn, looking for breakfast. The time was just before 7 A.M.

Molloy was a local boy made good. A native of Sommerville, not far from Harrisburg, his high school education and booming voice had made him news director at a Harrisburg radio station and, in 1974, he was named head of Dauphin County Civil Defense by Governor Milton Shapp.

The telephone rang. "I have a very small clock in my kitchen on the stove," Molloy said, "and I think it was in the vicinity of five to ten after seven in the morning when Maggie Reilly called."

Maggie Reilly told Molloy about the on-site incident at Three Mile Island.

"I don't know really what's coming off," she said, "but it doesn't sound wonderful and you might want to go to work early."

Molloy asked if an evacuation was called for and she said "No."

He didn't ask for any details of the accident itself. These didn't make much difference to Kevin Molloy. "The way the system is set up, we do not get any technical information," he said. "We are not necessarily interested in a whole lot of technical information. We are concerned with one thing: Must we evacuate? And the other important thing, if we do have to evacuate, obviously is which area and which way is the wind blowing?"

The smells of breakfast began to fill the kitchen. Molloy was half dressed, talking on the phone to Maggie Reilly when the fire monitor in his kitchen began to call his name. It was the county police and fire dispatcher, one of Molloy's men, asking him to call his office. Calling in, Molloy spoke to dispatcher Rick Dietrich. "Have they gotten in touch with you yet?" Dietrich asked Molloy.

"Who, Maggie Reilly? Yes, I talked to her," Molloy said. "Don't do anything yet. You didn't get a call from Three Mile Island, did you?"

"Yeah, a guy named Ron Warren called and he is declaring a slight emergency," the dispatcher responded.

"A site emergency?"

"No, a slight emergency. S-l-i-g-h-t."

Molloy quickly dressed and headed for his office in the basement of a county building in Harrisburg just a few blocks from the state Civil Defense office—PEMA. On the way, he drove in sight of Three Mile Island. "Other than the lack of steam coming from the cooling towers, it didn't look any different to me," he said.

The situation began to worsen even as Molloy was driving into Harrisburg. As he was coming up Second Street into the parking lot, a call came in over the radio that TMI had changed their condition to a general emergency. "We didn't really know what that meant, but it sounded worse than just a site or a slight emergency," Molloy said later.

Once in the communication center where most of the county's police and fire units were controlled, Molloy set about making sure state and county agencies were aware of the problem at TMI. He called PEMA for an update and ordered his dispatchers to call police and fire companies throughout the county as well as the Red Cross and state police. It was about then that a third call came in from Three Mile Island.

"Kevin, we're in this for real," said Dick Dubiel, the head radiological officer at TMI. "We've got the core covered right now, so I don't think we have a real big problem. But we've got some bad radiation readings that could in fact be erroneous, but we can't rely on that."

Molloy said he'd have Maggie Reilly call Dubiel at TMI. This was the last call Kevin Molloy and the Dauphin County Office of Emergency Preparedness would receive from Three Mile Island during the accident. Molloy continued to mobilize his forces and wait for word from TMI or from the governor about what to do. Word never came.

Dubiel was wrong in thinking the core was covered. At the moment he and Molloy were speaking, the reactor core was projecting eight feet out of its cooling water, bathed in steam and beginning to melt. Moreover, this conversation, recorded on the dispatcher's 911 emergency line recorder, showed that some in the control room suspected core damage hours before the utility would publicly admit it.

At 7:07 A.M., operators at TMI made their last call of the morning required by the plant's emergency notification plan. The call was to the Nuclear Regulatory Commission Region One office at King of Prussia, Pennsylvania, near Philadelphia. The NRC office was closed. An answering service took the emergency call, but it would be nearly 8 A.M. before anyone from the NRC would know about the accident that had begun almost four hours before.

When Met Ed finally did get in touch with NRC Region One headquarters—on their third attempt—the auxiliary operator spoke with Boyce Grier, NRC regional head. Grier did not waste time. He immediately dispatched a car with an investigator, a reactor inspector and a radiological survey team to the plant. Five minutes later, he sent off a second car, siren and lights blazing, for TMI. Then he called in the region's mobile lab from Connecticut. It might just be needed.

The station emergency plan did not require a call to NRC headquarters in Washington, D.C. The regional NRC office took care of that and began to contact the other federal agencies that participate in the Interagency Radiological Assistance Plan.

Boyce Grier's call to NRC headquarters in Bethesda set in motion an emergency plan that was well-tested. A general emergency called for activation of the agency's Incident Response Center, a suite of rooms equipped with communication equipment and computers for the use of NRC decision-makers as accidents progressed. And the inner sanctum of the IRC was a large, glass-walled conference room for the Executive Management Team—the men of the Operating Reactor Division who would be calling the shots.

The NRC was sprawled in office buildings throughout Bethesda; one office here, another there, leftovers from the days of the Atomic Energy Commission in 1974. It was necessary to have a common

34

place for the technical staff to meet under accident conditions, and it was in response to this obvious need that the IRC was established after the Fort St. Vrain reactor accident in 1978. And despite the industry's safety record, the IRC was used at least twice before Three Mile Island.

The last time the NRC staff had been called into the IRC was in the month before the accident at Three Mile Island. Then there was no problem with a reactor—nothing the NRC was used to dealing with.

A building subcontractor at the General Electric nuclear fuel processing facility in Wilmington, North Carolina, had stolen two shipping casks of slightly enriched uranium reactor fuel—about 150 pounds in all—and was demanding $100,000 for its safe return.

A sample of the stolen fuel appeared one morning on the desk of Wilmington *Star* managing editor Bill Smith. "It was a 35-millimeter film cannister full of stuff that looked just like cat litter," Smith said. "A note with a name and phone number was attached and the can was labeled with a skull and crossbones and the word 'radioactive.' I thought it was a crank."

It wasn't a crank. When Smith finally called the number on the note, it rang in the public relations office at the G.E. plant. Their system for keeping track of "every gram of radioactive material" had already detected the fuel loss, though the G.E. representatives were unable to say just how much material had been in the two shipping casks.

The IRC was activated, FBI informed, briefings were held for Congressman Morris Udall, Senator Gary Hart and House Speaker Tip O'Neill. The demand was taken very seriously in Washington, and most officials were especially sensitive about the incident's happening just as the Vice Premier of China, Deng Xiaoping, was visiting the nation's capital.

But there was no threat to the water system of Washington, D.C. The extortionist threatened instead to send sample fuel to the Clamshell Alliance and Ralph Nader's Critical Mass antinuclear groups.

Smith agreed to keep the story out of his paper until the thief was caught "to prevent a public panic and in exchange for the promise of an exclusive story that I didn't have a hope in hell I would actually get."

Within five days Smith was present at the arrest and the recovery of the stolen fuel. He did get his exclusive story. After this, the NRC Incident Response Center was closed down—until Three Mile Island.

When the IRC was activated again around 9 A.M. on March 28, a group of up to seventy specialists came together to provide advice to the utility—the licensee—and to see that the public health was protected. They communicated with the accident site and with experts around the country over twenty-six telephone lines. But only one of these lines reached the control room of Unit 2 and it was not established until after noon on Wednesday, twenty of these telephone lines were recorded and comprise the best documentation of NRC's thinking in Bethesda during the accident.

While the day-to-day operations of the NRC were conducted from Bethesda, the five commissioners who led the agency operated from offices on H Street, just a couple of blocks from the White House. The commissioners, too, would go into continuous session before long, giving the NRC three voices—the site, Bethesda and H Street —all saying different things.

By 8:30 A.M., the Department of Energy, Environmental Protection Agency, Federal Preparedness Agency and Department of Health, Education and Welfare had been informed. The Defense Civil Preparedness Agency learned of the accident from the Associated Press. The Federal Disaster Assistance Administration heard about the accident on the radio.

The DCPA and the NRC both contacted the White House situation room before 8:30 A.M. Operators in the communication center buried in a sub-basement of the executive mansion recorded entries for both calls in their log.

IRAP—the Interagency Radiological Assistance Program—was activated, and a radiological monitoring team from the Department of Energy's Brookhaven National Laboratory on Long Island set out for TMI. A special mobile laboratory from the Environmental Protection Agency's Office of Radiation Programs hit the road to provide backup for the Brookhaven effort. For purely political reasons, this mobile lab would never arrive at TMI.

Those federal agencies not directly involved in implementation of the IRAP plan began to send "observers" to Three Mile Island. There were soon more federal observers in Harrisburg than the Commonwealth of Pennsylvania knew what to do with.

Jessica Tuchman Matthews of the National Security Council staff received a call at 9 A.M. from NRC Commissioner Victor Gilinsky, who told her of the events at TMI and said that it "seemed to be an incident of some moment." Matthews was responsible for nuclear issues at the NSC and Gilinsky had felt that "somebody in the White House should be aware that this had happened."

Matthews, daughter of author Barbara Tuchman, quickly wrote a memo about the accident for her boss, presidential national security adviser Zbigniew Brzezinski, and took it in to him.

Brzezinski read the memo in his book-cluttered office in the Old Executive Office Building next to the White House. Matthews told him that news of the accident was already being spread by the media and that it had the potential of becoming a major incident. Grimly, Brzezinski took the memo over to the White House to show the President.

When Leslie Jackson heard about the accident at Three Mile Island, he quickly put all his resources on alert. As head of the York County Office of Emergency Preparedness, he was responsible for the safety of people in Goldsboro, the town nearest to TMI, and he wanted to be ready in case those people had to be moved.

Among the first of Jackson's calls was one to Ed Wickenheiser, news director at WSBA radio in York. WSBA was the local emergency broadcast station and would be needed to announce any evacuation.

Wickenheiser was a stocky fellow with a brisk, almost military, manner—not surprising for a member of the Marine Corps Reserve.

"Les told me about the accident, that there had been a cooling problem and that the plant was shut down, when he called around 7:30 A.M.," Wickenheiser said. "Of course, we didn't say anything about it [the accident] on the air." Ed Wickenheiser was the first newsman to learn of the accident at Three Mile Island, but he didn't tell anyone about it.

4

Captain Dave

"Captain Dave" Edwards drove a yellow Camaro through the early-morning streets of Harrisburg. He broadcast the traffic report for WKBO radio, surrounded in the cockpit by FM, CB and police radios. Whenever there was a car accident or a problem with traffic, Captain Dave reported it live to the people of "WKBO-land." Bigger cities might use a helicopter for the same purpose, but Dave liked his Camaro.

Early Wednesday morning, traffic was moving smoothly into Harrisburg and Captain Dave cruised Interstate 83 listening to police calls on his scanner. About 8 A.M., a call came in on the CB from one of Dave's good buddies who mans the Middletown REACT frequency during that early hour.

"He said something about the Colonial Park Fire Company being put on alert," Dave said later. "Nothing about a reason, just that they were on alert and that it was pretty unusual and did I know anything about it? Well, there was nothing coming across the scanner, so I called it in to the station and told Mike."

Mike Pintek was the news director at WKBO radio. He and his wife had lived all their lives in Dauphin County. Pintek tried to insert an hourly five minutes of news somewhere inside the top-forty format of WKBO. He wanted to move up in the news business, but not too far. Pintek liked living in the country. He liked driving the yellow Camaro with "WKBO" painted on the side. "If I drove a car like this in New York or Philly," he said, "nobody'd even look up."

39

Pintek had just finished the 7:55 A.M. newscast when Captain Dave's radio call came in. He took the call in the cramped 12-by-12 newsroom and went to work on the telephone to check out the story. "Nothing was on our scanners yet," Pintek said, "so I called Dauphin County Control, where they handle the police and fire calls for most of the county." The dispatcher denied any activity in Middletown, so Pintek called Captain Dave back to let him know the story wasn't correct. And Captain Dave said, "You know, I didn't see any steam coming out of Three Mile Island this morning."

Like most reporters in the Harrisburg area, Pintek hadn't had much contact with Three Mile Island other than to cover a few rate-increase hearings and to take a tour of the plant two years before. On that tour, he'd picked up the business card of then plant manager Jack Herbein. On a hunch, Pintek decided to call the number on the card.

To Pintek, the TMI switchboard operator sounded confused, even frightened. When Mike Pintek called, he asked for public relations, but that's not what the operator gave him. She connected him with the control room of Unit 2.

"A guy answered the phone and he was very excited, almost in panic. When I visited the place two years before, everybody seemed so calm it was eerie, but this fellow was anything but calm," Pintek said.

The control-room technician who answered Pintek's call didn't have time to answer questions from the press just then. "I can't talk now, we've got a problem," he said. Remembering the call for firemen, Pintek asked if they had a fire at the power plant. "Fire engines? No, I don't think there are any fire engines," the man replied, misunderstanding the question. "I can't talk now, we've got a problem. Call Reading and talk to them." Then he hung up.

Pintek called Metropolitan Edison headquarters in Reading and asked for Blain Fabian, chief public relations officer for the utility that owned the Three Mile Island nuclear plant. Fabian was in a meeting with his entire staff of communication services. "When I heard they were all in a meeting at 8:07 A.M., I knew something was wrong," Pintek said. "I told them to get him out of the damn meeting."

Fabian came on the line and said there was a general emergency. He said that it was just a formality required by the NRC when certain conditions exist. What conditions? There was a problem with a feedwater pump. The plant was shut down. There was no danger off-site. No danger to the general public.

That was the first story Mike Pintek and WKBO went with for their 8:25 A.M. newscast when they were the first to tell the world that the worst nuclear accident in history had begun: There is no danger to the general public.

When word of the accident at Three Mile Island was released by WKBO and, later, by the wire services, news organizations throughout the nation began to respond to what still seemed to them to be a minor accident near Harrisburg. They began to mobilize reporters and photographers to cover the story. Most of these journalists were assigned to the TMI story because they were available or nearby. The science specialists were sent in later.

Don Janson was sent to cover the story for the *New York Times* because he was available and based in the Philadelphia bureau. Gary Shepard was the first CBS correspondent to go to Three Mile Island simply because he was the first CBS correspondent to get to work in New York that Wednesday morning. The same fate awaited Bryce Nelson, who went in earlier than usual to the Los Angeles *Times*'s Washington bureau. But many of the first reporters to cover the accident were political writers assigned to the Pennsylvania state capitol nine miles away.

The early wire reports were sketchy at best and came under harsh criticism from the more experienced science writers. Roger Witherspoon of the Atlanta *Constitution* found technical errors in the first Associated Press piece he saw about the accident, when his wire editor came to ask him, Just how important is this nuclear accident anyway? "Finally I had to say don't accept any wire stuff out of Harrisburg. The AP didn't know what they were doing."

41

In contrast, Jonathon Ward, an associate producer for the CBS Evening News, found the early wire stories very enlightening, if suspicious.

"A science specialist in any large news organization has to have a highly developed 'crap detector,' " Ward said. "Most of the people I work with don't know what is a good science story and what isn't. They sometimes ignore really important stories, not realizing that they are important but, more often, they want to declare that the end of the world is upon us for what turn out to be relatively minor problems at a nuclear power plant, for example. That's where the crap detector comes in. I have to decide for them in a minute what is and isn't a real story."

Ward's crap detector was put to work shortly after he arrived at the CBS News Broadcast Center in New York on March 28. Cindy Samuels, the assistant Northeast bureau chief, showed him an early wire story about the accident. "It says there's been an uncontrolled release and a general emergency," she said. "What does that mean?"

"It certainly sounded bad," Ward said later, "but the wires can blow these things up a bit and I really wasn't sure whether there was a story there or not."

Ward told Samuels to "get someone there but don't be surprised if it doesn't turn out to be a real story."

"This is one of the times my crap detector was in error," Ward said. "But we had another story not long before then when a case of radioactive cobalt supposedly fell off a truck somewhere. We sent out a helicopter and a crew only to learn that it wasn't radioactive cobalt at all. I'm suspicious of nuclear accident stories."

Jonathon Ward was probably the broadcaster best prepared for Three Mile Island. He had done many nuclear stories before, knew the controversial subjects of waste disposal and low-level radiation. He also had large files of background material on the nuclear industry.

"Bob Schakne and I had been making up files on every nuclear plant in the country, with pamphlets, clippings and a color slide of each. We were waiting for the big nuclear accident story we figured would come sooner or later." Unfortunately, Ward knew nothing about TMI other than its name.

Though he doubted the seriousness of the accident, Ward had little choice but to send in a crew to check it out. In the past, CBS had been beaten to important stories because it had not been quick enough in sending reporters to the scene.

"There has been a problem of our starting late on stories of lesser significance," Ward admitted. "NBC flies in with two crews; CBS has one crew, one correspondent and one editor; ABC has no one or three crews—either one is wrong." Stories where CBS had been slow in reacting included the fall of the Shah's government in Iran and the collision of two Boeing 747 airliners at Tenerife in the Canary Islands.

"We really were burned on the Tenerife crash story," said Ward. "It was a long way to send a chartered plane for a story that might not be as big as it first sounded. We guessed wrong and, since then, we've compared other stories to Tenerife. It's become a joke at CBS —whether a story is another Tenerife or just a 'Ninerife.' "

True to form, the CBS Northeast bureau sent a chartered aircraft with one crew, one correspondent and one editor to Harrisburg late on Wednesday morning. The correspondent was Gary Shepard.

Many newspapers tried to cover the story by telephone or elected to wait and see if the accident was really significant. The Philadelphia *Inquirer* elected on Wednesday to use just the wire-service accounts of the accident. A BBC correspondent flew in to Harrisburg by airline early Wednesday afternoon, decided there wasn't a good story at TMI and returned to Washington, D.C. He had to return in a hurry the next day by chartered aircraft.

In all, perhaps two hundred reporters were on their way to Three Mile Island by late morning. By then, one local paper was printing its third story about the accident for that day.

Like many others, Mary Bradley of the Harrisburg *Evening News* was assigned to cover the story of the accident because she had some free time. "I'd just finished up another story before our 8:30 A.M.

deadline for the first edition," she said later. "One of the other reporters had heard about the accident on his car radio while driving in to work. I was free, so Pat Carroll, the city editor, gave it to me."

Bradley, slim and tall, had been working in the main office of the *Evening News* for two years. Prior to that, she'd been assigned to the bureau in York, twenty miles south of Harrisburg.

Her first action was to call TMI directly, but she had no success in getting past the plant telephone operator who suggested she call the utility's main office in Reading. She had little luck in calling the Reading office either, since all of the staff of communication services —Met Ed's public relations office—were in a meeting and none of them would come to the phone.

Finally Bradley called PEMA to see what the state Civil Defense office knew about the accident. She spoke to John Comey, PEMA's only public information officer, who told her more than either of them realized.

"He gave me some sketchy information about the accident—when it had happened, that the plant wasn't generating electricity and that there was a general emergency. Though he couldn't tell me what a general emergency was, Comey said the plant had shut down because it had 'failed to fuel.' Neither of us really knew what that meant, and I wasn't able to confirm it from another source, so we left it out of the first-edition story."

It was obvious that the accident was an important story, so the first-edition deadline of the *Evening News* was pushed back to include Mary Bradley's first piece about TMI. She wrote with city editor Pat Carroll looking over her shoulder, pushing her to complete the story by 9 A.M. That 9 A.M. deadline was the last deadline the *Evening News* would have for the next six days as the story of the accident evolved and the paper went to continuous remakes of the front page.

What Mary Bradley and John Comey didn't know was that "failure to fuel," the presence of damaged fuel in the reactor, was an important indicator of the seriousness of the accident. It meant that at least some of the fuel cladding had been damaged and perhaps that parts of the core had melted. It was also a fact that Metropolitan Edison would publicly deny for the next twenty-four hours.

The cooling towers of Three Mile Island were just out of sight of Middletown, Pennsylvania. Even out of sight, TMI had a strong presence in Middletown. The plant contributed to the obvious prosperity of the sleepy town, with its treelined streets and frame and brick houses from the last century, some of them—on the less expensive side of town—covered in asphalt-based ersatz brick siding. Many Middletown men worked on the Island and the seemingly endless construction of first Unit 1 and then Unit 2—between them they had taken more than ten years to build—had brought millions of dollars into the local economy.

The utility had little trouble getting along with the local population. During licensing hearings for Unit 1, the greatest topic of local interest was whether Met Ed would build a river recreation area on an unused part of the island. Only a single local resident objected.

"Picnic tables!" John Garver had screamed. "They're raping the damn river and they're going to put up some lousy damn forty-dollar *picnic tables!* They're going to dump their filthy shit in the air and the water and give us *picnic tables!*"

John Garver's objections were not taken very seriously by the people of Middletown.

But TMI wasn't the only business going on near Middletown. Just to the north, the Fruehauf trailer factory—largest in the world— employed nine hundred in its cavernous buildings, booming with riveters and hydraulic sheet-metal presses. And just beyond Fruehauf, in Steelton, was the nineteenth-century Bethlehem steel-rolling mill, covered with grime and dust, but helping to make the local economy work.

And the economy did work. Life was steady in Middletown and like almost everyone else in town, except some of the technicians at TMI, Mayor Robert Reid was a product of Middletown. Thirty years ago, he was a 132-pound fullback on the Middletown High Blue Raiders football team. A feisty, pugnacious black man, Reid was proud of his high school and college football career, of being a college boxing champ and of being the $150-per-month mayor of this town of 11,000.

When Reid played for the Blue Raiders, his nickname was Bird. Now he taught government in the same school and his son was the football player in the family. A member of the Middletown Blue Raiders, young Reid's nickname was also Bird. Not much changes in Middletown.

Young people seemed loath to leave it, even for the challenge of college or the excitement of the big city. Families just stayed on, generation after generation, saying, "I was born in Middletown, I live here and I'll die here." Some of these families had been saying just this since 1755.

Early each weekday, seven or eight local men storekeepers, local businessmen and teachers would meet for coffee or breakfast in Kuppy's red brick diner across from the Middletown Post Office. Robert Reid, the only black mayor in Pennsylvania, was a member of the regulars at Kuppy's. And while Unit 2 was "burping" radiation three miles away, Robert Reid and the others were having a second cup of coffee.

Mayor Reid went to work knowing nothing of the problems at TMI. In the middle of his first-period American Government class, though, word came in from the town's Civil Defense director, Butch Ryan, that something was happening on the Island. Reid left his class in the hands of the assistant principal and went to his office downtown in the municipal building.

"All that I learned when I got to the office was that there had been an on-site emergency declared," he said. "So we sat there and we listened to the television. We changed from channel to channel, and each channel gave us different information. We tried the radio and each station gave us different information. So I told him, I said, 'Butch, we got a problem, we don't know what's going on.'

"Because people were calling us who had relatives working down there and people just wanted to know what was going on. We couldn't give them any answers. We told them to call the county— we never did find out whether those people got through or not, the county's phones were so tied up. I guess it was about eleven o'clock I told Butch I was going to call the Island.

"I got an operator, she couldn't tell me anything. She did tell me to call Mr. Fabian over at the Reading office. I called there and he was in a meeting and he couldn't talk to me and eventually a Mr. Guerin called me back. He was in Reading at the home office and he assured me that no nuclear particles had escaped.

"No particles had escaped, no one injured and I said, 'Hey, that's great!' So I went back to work. I turned on my car radio and the announcer said that particles did escape.

"Mr. Guerin called me again later that day from the Reading office and told me he wanted to update our conversation. I said, 'What do you want to tell me—that nuclear particles had escaped?' He said 'Yeah.' I knew we were in for a heck of a time when the announcer told me that particles had escaped and I had just talked to him about twenty seconds before that and he said no particles.

"And I said, 'God almighty, here it is and we don't even have a plan.' Caught me short, caught me with my pants down without an evacuation plan."

5

A General Emergency

The situation continued to be tense in the control room of Three Mile Island's Unit 2. Safety director Gary Miller arrived and, by 7:35 A.M., had decided to change the status of the accident from a site emergency to a general emergency. Though the utility continued to reassure public agencies that there was no danger, a look through the Met Ed emergency plan would have shown that a general emergency is "an incident which has the potential for serious radiological consequences to the health and safety of the general public." There is no level of emergency in the nuclear industry beyond a general emergency. It is the worst. But nobody from Met Ed would mention that fact to the press or public on Wednesday.

And if there was no need for an evacuation, that was just as well, since none of the communities within five miles of the plant had evacuation plans.

At 7:35 A.M. PEMA operations officer Dick Lamison took the call from Gary Miller announcing the change to a general emergency. Lamison's log entry notes "a change in alert status to general alert situation" and that the reactor had tripped because it "failed to fuel." It would be two days before Met Ed would talk publicly about the possibility of the core's having been uncovered or about extensive fuel failure, yet both conditions seemed likely to Miller and Dubiel before 8 A.M. that Wednesday.

By now the day shift workers were beginning to arrive at the plant. Miller decided that nonessential personnel should stay off the Island

48

and so had them report to the observation center for a head count. The observation center was a small brick building where tourists could view the power plant across the river while seeing films about the value and safety of nuclear power. Nobody was watching the films that morning.

With nothing much to do at the observation center, a couple of hundred Met Ed employees stood around, talking and smoking, while a few more energetic souls threw a Frisbee. Bill Parker took charge of making sure which employees were at the observation center and that none had gone onto the Island before the gates had closed.

Terry Mackey, the plant's supervisor of quality control, arrived at the locked North gate around 7:10 A.M. Mackey had been the head mechanical engineer for the startup of Unit 2 and felt in a sense disappointed that his new position placed him among the nonessential personnel waiting at the observation center. He would really rather have been in the control room. In fact, after checking to see that all the people in his department were safe, Mackey did go into the Unit 2 control room to see what was happening.

"I just told the guards, 'I've got responsibilities,' " Mackey said. " 'How am I going to find out what's going on? I've got to go in.' "

"So I went on up to the control room and didn't necessarily talk to anyone in particular. I just kind of walked around. It was three hours after the trip and things were generally calm. Gary Miller and the superintendents and the off supervisors were all in the shift supervisor's office, apparently discussing the situation.

"Down near my office in the service building is the auditorium where they were holding everybody from the third shift, eleven to seven-thirty in the morning. They were holding everybody on-site in the event that they wanted to talk to those people later."

Leaving the Island, Mackey went back to the observation center to report to George Troffer, his boss at Met Ed headquarters in Reading, who did not yet know that a general emergency had been declared.

"What's going on?" Troffer asked. "We're getting bits and pieces down here about something's wrong."

49

Mackey told Troffer what he knew of the situation in Unit 2. He agreed to stay at the observation center, calling in to the control room four times an hour for an update to give to Troffer in Reading. This is how much of the Met Ed management got its early information about the accident: through Troffer and Mackey and a pay phone at the observation center. It's also how the press got some of its best information. Troffer soon became the most open and reliable source for the press in the utility.

The headquarters of Metropolitan Edison were in Reading, Pennsylvania, right in the middle of the area served by the utility. There, thirty-five miles east of TMI, Walter Creitz ran his third of the General Public Utilities empire.

Creitz began his career with Met Ed in 1948 as an engineering cadet. His entire working life had been spent with the Pennsylvania utility. Since he became president in 1972 Met Ed had turned into virtually an extension of Walter Creitz's personality. And on the morning of March 28, that personality was mainly concerned with how to deal with the press.

Word of the accident at TMI did not reach Reading until after seven, when Richard Klingaman, one of the engineers based in Reading, received a call from TMI. Klingaman reported the accident to Creitz's secretary and to his boss, George Troffer, head of quality assurance for Met Ed. Troffer knew immediately that it was a bad accident. His reaction to news of the high containment radiation readings was "instant concern—like having a pistol put in my face."

With Met Ed's vice president for Electrical Generation Jack Herbein in Philadelphia on Navy Reserve duty, Troffer called Robert Arnold, General Public Utilities' vice president for Generation, in Parsippany, New Jersey. Arnold dictated an official statement to be released to the press after 9 A.M. Arnold's technical statement did not refer to release of off-site radiation or to the declaration of a general emergency, but that was of little import. The statement was ignored anyway.

When Met Ed communication services chief Blain Fabian—supposedly the chief spokesman for the utility—delivered Arnold's statement to the public information officers who manned Met Ed's telephones in Reading, it was not used. At least two members of the communication services staff did not read the statement to the press or public. As Judy Botvin, a member of Fabian's staff working the phones that morning, explained her decision not to read Arnold's release: "I didn't understand it. I didn't feel I could explain it to anyone."

For most of the morning, Botvin, Dave Klusick and Don Curry tried to answer the questions of the press with a very short statement prepared over the phone by Herbein and Fabian:

"The nuclear reactor at Three Mile Island Unit 2 was shut down as prescribed when a malfunction related to a feedwater pump occurred about four A.M. Wednesday. The entire unit was systematically shut down and will be out of service for about a week while equipment is checked and repairs made."

The optimistic message conflicted with what Fabian was told after 8:30 A.M. when he talked to George Kunder in the Unit 2 control room. Kunder was talking about core damage and about external radiation releases—both much worse than what Fabian's people were claiming on the telephone. A cautious man who'd spent seventeen years with the utility in the same job, Fabian told his people not to repeat what Kunder had said: the news had not come as it should from station manager Gary Miller and so was "unofficial" and had to be confirmed. While Fabian went off to try and confirm Kunder's pessimistic message, Reading continued to insist that there was only a minor problem at TMI.

As wire-service reports of the accident went out, calls from reporters all over the country began to flood the Reading offices. Later that morning accountants and secretaries were added to the staff dealing with the press. The additional amateur manpower was not well accepted by the press.

"People were enraged that we didn't know everything," Judy Botvin said later. "Reporters were very indignant. They would say, 'Well, if you don't know, what is this? Met Ed doesn't know?' I would say, 'Well, I'm not saying Met Ed doesn't know. I'm saying that this is all the information we have, and that the people who have to put answers together are busy with the emergency at the plant.' 'Well, let me talk to Jack Herbein.' 'I'm sorry, I can't put you through. He is busy.' And they didn't want to accept that. Each one wanted to talk to Jack Herbein."

Public information dominated activity in Reading for Met Ed. Troffer was put to work trying to answer the questions of the press, as were other technical people at the headquarters. Yet the information available in Reading was minimal—there just wasn't that much the untrained public information officers could say with the sketchy information coming occasionally from the control room.

But they tried to answer every inquiry, even if the answers were woefully inadequate. They tried because Walter Creitz was president of Met Ed and Walter Creitz was concerned. "It was only several days before that, that I had seen a movie called *The China Syndrome,*" Creitz said later, "and I was particularly sensitive to having the ability to tell the public what happened."

Bill Gross was normally in charge of the Met Ed observation center. A former English teacher at Middletown High, who admitted that he became a teacher to coach basketball, Gross made slide presentations in area schools and spoke to local groups about nuclear power. Driving to work just after 7 A.M., Gross found the gate to TMI locked. Further down the road, he found two to three hundred employees milling around the lawn of "his" observation center. He also found himself the only public voice of Metropolitan Edison.

"When I parked the car, I talked to a couple of girls—secretaries—sitting in their car, listening to the radio, and I said 'What's going on?' They didn't know—they were trying to find out from the car radio. My initial thought was that it was a labor problem. So I went into the building and there were seventy-five to a hundred people there, all demanding to know where I kept the coffee."

Among the employees waiting at the observation center was a pregnant secretary. She asked Gross if it was safe for her to remain there, a quarter-mile from the plant. Gross and one of the plant health physicists told her not to worry.

Calls from the press began to come in before 8 A.M. The observation center had a single telephone and it was soon ringing constantly with calls from all over the country. The callers were asking for information that Bill Gross couldn't provide—he wasn't able to get his own call into the control room of Unit 2 for information to give the press or employees waiting around him.

It would be 9:30 A.M. before Bill Gross could speak to Gary Miller in the Unit 2 control room less than a half-mile away. Then another half-hour would be spent writing a statement and checking it with the Met Ed main office in Reading. At 10 A.M., Bill Gross, the former basketball coach from Middletown High, would make Metropolitan Edison's first public statement about the accident:

"We had a turbine trip early this morning due to a feedwater problem in the secondary side of the plant—not a nuclear problem. This caused the reactor to trip on high pressure, which was followed by the pressurizer relief valve relieving, which resulted in a radioactive water release in the reactor building.

"Since this radioactive cold water was released inside the reactor building, this led to the emergency plan implementation. Radiation monitoring teams have been dispatched on- and off-site to monitor for possible external radioactive releases. None has been found so far and we do not expect any. We are presently bringing the plant down to an orderly cold shutdown with no consequences to the public expected."

Back inside the control room of Unit 2 there was plenty to do. Gary Miller had declared the general emergency based on a reading of more than 8 rems per hour measured by an instrument mounted on the reactor containment dome. This reading corresponded to 800 rems per hour when corrected for shielding around the detector. From this reading, the Met Ed health physics staff calculated the exposure to the nearest population, using windspeed, direction and weather conditions. According to Tom Gerusky, director of the state's Bureau of Radiological Protection, ". . . it was calculated and given to us that the dose rate directly west of the plant across the river would be ten rems per hour from noble [inert] gases. That is the information we got over the telephone."

Ten rems per hour is a lot of radiation. Workers at Three Mile

53

Island were normally allowed to receive no more than 5 rems per *year*. Ordinarily levels near the plant were on the order of 100 *millirems* per year—one tenth of a rem. So the expected radiation dose in Goldsboro, across the river from the plant, was over 875,000 times the normal radiation level for that area.

Maggie Reilly of BRP relayed the information to PEMA and suggested that preparations be made for a possible evacuation of Brunner Island and Goldsboro, just across the river from the plant. She also sent a helicopter to Goldsboro to verify the 10-rem figure.

"The figure was ginned up by the utility in response to their emergency plan," Reilly said. "The plan postulated a 0.2 percent per day leak rate from the containment and we weren't sure there was pressure in the containment to make it leak that much."

Within a few minutes, the radiation monitoring team reported no significant readings in Goldsboro. The reactor containment was at less than atmospheric pressure, so none of the feared noble gases were leaking out of the structure. But that situation could still change, and quickly. "We'd told PEMA to tweak the county [inform them of the danger]," Reilly said, "and now we had to go right back and tell them to detweak."

The concern about radiation remained, and the worst recognized danger was the possible presence of radioactive iodine 131 concentrated in local dairy products. Maggie Reilly began to contact other state departments about this problem.

"I got in contact with Agriculture, who are always high on the list because it's hard to conceive of a reactor accident where you're not going to have a cow problem, which is something that amazed me throughout this whole thing; I just couldn't imagine how come we didn't have any iodines. It just ruined my whole day."

By 7:45 A.M. PEMA Director Oran K. Henderson decided it was time to inform Governor Richard Thornburgh. Henderson called him at the Governor's Mansion, where Thornburgh was getting ready for breakfast with a group of Democratic freshman legislators. The governor was also new in Harrisburg, having taken office only two months before. Thornburgh had served before as a district attorney, U.S. attorney, and U.S. Assistant Attorney-General. A Republican governor in a Democratic state, he needed the support of at least some of those freshmen.

The two men spoke for only a moment. They had met just once before. Thornburgh asked Henderson to notify Lieutenant Governor William Scranton, chairman of the Pennsylvania Emergency Management Council and directly responsible for state response in such emergencies. And just before going in to the working breakfast, Thornburgh called his press secretary, Paul Critchlow, to find out what Critchlow knew about "some sort of problem at TMI."

With a real emergency in hand, Pennsylvania Emergency Management Agency Director Oran Henderson came into his own. It was almost as though he had been waiting for an accident and a chance to take command.

Until late 1978 PEMA was called the Pennsylvania Office of Civil Defense. But by 1978, it was clear that the lack of an impending war with the USSR, and the fact that most Civil Defense activities in Pennsylvania involved planning for and responding to floods, made a name change logical. The PEMA Director said, "Civil Defense had developed an outmoded image and we wanted to make ourselves a more active and acceptable agency in the eyes of the public."

In reality, though, PEMA was still Civil Defense, still had a library of films to show schoolchildren how to protect themselves during a nuclear attack, and was still housed in the basement in a "secure" location—secure from specific levels of radiation spewed from whatever Soviet bomb or missile happened to fall on the Fruehauf trailer plant or the Hershey chocolate factory.

Reporters and even some state workers came to call the PEMA basement command post "the bunker." Only it wasn't really a bunker at all. It was just ugly and very military-looking, with little glassed-in cubicles for offices. Gray desks filled the windowless rooms and not much color was evident, except for the orange of the lower four feet of the walls and the maps of Three Mile Island mounted in each room. Each map had a big dot in the center, where the power plant is, and concentric circles at 5, 10, and 20 miles from the dot. Some maps eventually had hundreds of red and blue pins, some with flags on them, to indicate evacuation routes.

The fact that PEMA headquarters had a military look should have been no surprise. PEMA, like most other states' departments that used to be called Civil Defense, was staffed mainly with ex-military men, enjoying one pension while qualifying for the next. And many of them were colonels before making the switch.

"It's almost impossible to deal with these guys," said one PEMA staffer. "They have this military mentality that doesn't have much to do with intelligence. And if you disagree with them in a meeting or say something they don't like, they start to tap their rings loudly on the table—class rings from West Point—just to show who's 'made it' and who hasn't."

It was this "military mentality" that produced the PEMA plan for evacuation of strategic parts of Pennsylvania during a nuclear war. The plan calls for moving 8.5 million people from their homes in areas of danger and generally putting them in mass groups in other areas that are only a little less threatened. And military efficiency feels the 8.5 million people can all be moved in 32 hours. In contrast, the Harrisburg Police think it would take 20 hours to evacuate the state capital alone.

One of the PEMA colonels who doesn't tap his West Point ring on the table is the PEMA director, Colonel Oran Henderson. Henderson didn't go to West Point. He joined the Army as a private in 1939 and still has a preference for the old "Sergeant York" style of tin hat over more modern combat dress.

Henderson served as commander of a combat infantry brigade in Vietnam. He was known as an effective and strong leader with what the Army considered to be only a minor fault. According to Brigadier General Andy A. Lipscomb, a former superior of Henderson's in Vietnam, "I knew him to be a brave individual and, I thought, a fairly strong leader. I wasn't sure that he was necessarily the most intelligent of the people I had commanding the brigades."

Henderson probably would have left Vietnam a Brigadier General if he hadn't been the commander of Lieutenant William Calley in 1968 when Calley led his platoon in the massacre of more than one hundred villagers at My Lai.

My Lai destroyed Colonel Henderson's military career. He survived a court-martial, but found himself, still a colonel, commanding a backwater installation at Fort Indiantown Gap near Harrisburg. When the Civil Defense job came along in 1976, he leaped at the chance of starting a new career.

The PEMA director was at once a part of the military and apart from it. Having come up through the ranks, he annoyed the colonels who'd gone through "the Point," and could never become one of them. At the same time, he couldn't suppress all that thirty-seven years in the Army had done to him. His underlings in Vietnam found him to be a hard taskmaster. One of them said, "I was scared to death of Colonel Henderson . . . He's just got to be the hardest man I've ever worked for." And Henderson had some trouble at first in adapting to the different pace of a civilian agency like PEMA.

Shortly after taking over as director of what was then Civil Defense, Henderson met with a group of community Civil Defense directors—barbers, garage mechanics, farmers—all of them unpaid for their service. When Henderson came into the room filled with ninety talking men, each glad to be in Harrisburg for the day at state expense, he called out, "At ease!" There was a quick, hard silence and Oran Henderson suddenly knew what not to say to men whose last military experience had been at the invitation of a draft board in 1943.

6

The NRC Arrives

When the first NRC station wagon from King of Prussia rolled up to the gate of Three Mile Island at 10 A.M., inspector Jim Higgins noticed that "the plant was essentially deserted."

The five men checked in, receiving the same film dosimeters to measure radiation exposure that any other visitor would get. Perhaps these dosimeters got a bit more scrutiny at day's end, though.

First they went to the Unit 1 control room for a briefing. While the Unit 2 control room still controlled the stricken reactor, Unit 1 had become the center for radiological monitoring of the area around the Island. After a briefing by the Unit 1 shift supervisor, the NRC health physicists took their equipment out into the plant to measure local radiation, Charles Gallina stayed in the Unit 1 control room and Jim Higgins began the quarter-mile walk past the 300-foot cooling towers and between the turbine and auxiliary buildings to the control room of Unit 2. On his walk, Higgins passed utility workers already roping off areas of the plant that showed high radiation contamination. Some hot spots in high-traffic areas were marked with signs reading "DON'T WALK. RUN!" Some of these signs would remain in place for months.

Higgins arrived at the Unit 2 control room just as Gary Miller was sending nonessential personnel out of the facility. It wasn't that he felt they were in the way; rather, there just weren't enough respirators to go around.

Radioactive cooling water on the floor of the reactor auxiliary building was releasing fission products into the air and some of them had wafted into the control room with a change in wind direction, setting off radiation alarms inside the concrete structure.

"The system that normally takes iodides out of the control-room air was working," Gallina said later, "but we were not dealing solely with iodides. As a matter of fact, at the time there was no iodine, but the xenon gases were getting into the control room. Until they could identify the fact that iodine was not present, the operators had to go in respiratory protection."

Higgins entered the control room to find "twenty to thirty" operators, auxiliary operators and others working in respirator masks and having a hard time communicating with each other.

"You can see, however, talking is fairly difficult," Higgins said later of the respirator masks used that day. "It is difficult to hear and difficult to talk and make yourself understood, so that if I was trying to give this description while wearing a respirator it would be very difficult. I would have to yell and be asked to repeat and be asked to talk slowly."

At 10:30 A.M., the reactor was still "cooking"—sitting with the core uncovered and minimal cooling. Chick Gallina was dumfounded at the status of the reactor: ". . . we found the reactor in a state, you know, where we never—for want of a better word—had seen one in this state."

Higgins was even more concerned about the poor organization in the control room of Unit 2. "Mr. Miller was in overall charge, however there was not a clear, well-defined organization structure. The operators were out actually doing the manipulations at the panels based on direction from Mr. Miller or Mr. Logan, the unit superintendent. There were a lot of short-term problems that were being addressed, and as these problems came up, the people sort of caucused, you might say, discussed them, tried to get the best opinion of all people present, which included NRC opinions, from myself and other people that were there."

Miller had a more optimistic view of the control-room hierarchy and viewed himself in complete control. "I knew exactly what was going on," he said later. "I did not know how to get out of it, but

59

I had not been in this situation and did not imagine that I would be, but I had no panic. There was no panic in the control room. The direction was coming from me. I knew all the radiation readings. I knew I had a hell of a problem."

Gary Miller's background was a bit different from most of the technical staff at TMI. He was not directly a product of the Navy nuclear program. A graduate of the Merchant Marine Academy, his introduction to nuclear power came while working as a civilian at the Newport News shipyard in Virginia. Starting as a test engineer on nuclear submarines, he became chief submarine test engineer and ended his eight years at Newport News as construction manager for the nuclear aircraft carrier *Nimitz* and nuclear cruiser *Eisenhower.*

Miller went to Met Ed to run the acceptance test program for Unit 1 in 1973. By 1979 he was in direct charge of both reactors as the station manager, a job he'd inherited from Jack Herbein.

Higgins and Gallina established a phone line to NRC regional headquarters at King of Prussia. Until noon, Higgins tried to speak over the phone, through his respirator, to someone in the regional operations center who, in turn, relayed the same information on to the NRC Incident Response Center in Bethesda. It was not a very efficient or accurate way to transmit information for the experts in Bethesda to deal with. "There were times," Higgins said, "when I just had to take off the mask to talk on the phone for a minute or two. Then I'd put it right back on."

Miller had a difficult situation to deal with—much more difficult than he would admit. Radiation readings from the dome monitor had reached unbelievable levels. From the 800-rem-per-hour level that had triggered a general emergency at 7:35 A.M., the monitor indicated a level of 1,000 rems per hour by 8:40 A.M. and 6,000 rems per hour at 9 A.M. And it continued to rise. Radiation in the containment was so high that, should someone have been foolish enough to enter, he would have had only a few minutes to live.

"It [the dome monitor] was reading 40,000 or 50,000," Miller said. "I mean, that was beyond what I had ever envisioned seeing on the dome monitor . . ."

Miller recognized that there were steam bubbles in the cooling system, preventing normal circulation of primary-coolant water. But he also thought the core was still covered and was reluctant to start the main cooling pumps to avoid damaging them. "I might need those pumps later on," he explained.

The plan at 10:30 A.M. was to open the blocking valve that had finally closed off the malfunctioning Electromatic relief valve on the pressurizer. Opening the blocking valve would allow the system pressure to drop to a point where the automatic core-flooding tanks could dump 500,000 gallons of borated cooling water into the reactor vessel, lowering temperatures enough to activate the decay-heat-removal system that used Susquehanna River water to directly remove heat from the reactor.

Fred Scheimann cracked open the blocking valve at 10:30 A.M. and system pressure began to drop below 1,000 pounds per square inch, through 900 pounds, 800 pounds, 600. For six hours, the pressure dropped, never getting down to the 500-pound level required to trigger the half-million-gallon flood. The flood system was designed to be triggered by a large loss-of-coolant accident like a broken major coolant line and could not be triggered by the small pressurizer relief line.

The result of this de-pressurization effort was more damage to the reactor core, release of more radioactive decay products into what little cooling water remained in the core, more radioactive coolant in the bottom of the containment and a lot more hydrogen gas evolved from the zirconium-water reaction. Some of this gas rose to form a bubble in the top of the reactor vessel—a bubble that would cause great concern two days later. And some of the gas followed the steam path out through the pressurizer relief line and began to accumulate in the steamy atmosphere of the concrete containment dome.

Communication problems between Three Mile Island and the NRC experts in Bethesda remained difficult. Up until noon, they continued to be transmitted through a middle man in King of Prussia. The respirator masks that were put on or taken off with every change in wind direction hampered speech. The Unit 2 phone line failed several times, and information had to be sent over an intercom to the Unit 1 control room and then through the same difficult circuit to Bethesda. Finally radiation levels got so high in the Unit 1 control room that it had to be evacuated.

With a roomful of experts at Bethesda, the NRC had, it seemed, everything but accurate information to work with. By noon, a separate line from Unit 2 to Bethesda was set up, and an inspector stationed on each end to feed the experts information. But, by then, a far more important task occupied several of the NRC technical staff:

"Good afternoon, Hertz."

"Uh, yes. This is Bernard Weiss of the Nuclear Regulatory Commission in Bethesda. There is a congressional delegation traveling to Harrisburg this afternoon. They'll need several cars waiting at the Harrisburg airport. Can you help me?"

"I'm sorry, sir. We have no cars available in Harrisburg."

"Miss, this is an emergency government request."

"Sorry, sir."

"Good afternoon, Avis . . ."

The reactor continued to cook while the system pressure was allowed to drop. At 1:50 P.M., Gary Miller heard a sound "like a valve slamming under high pressure" somewhere in the reactor building. None of the NRC inspectors noticed the sound or recalled the event at all.

Ed Fredrick was watching the containment pressure meter, a graph recorder that inks a permanent record of the containment pressure on a turning roll of paper. They were trying to keep the containment pressure below 4 pounds per square inch to minimize radiation leakage through the containment walls and to keep the automatic isolation valves from closing off the containment as they would in a major accident, making it impossible to pump coolant from the containment sump and out of the building. The containment pressure was about 2 pounds per square inch.

Just when Miller heard his "slamming valve," the needle on the pressure recorder in front of Fredrick jumped instantly from 2 to 28 pounds per square inch. Fredrick pointed out the peak to Craig Faust and others who stood near him.

Fredrick said later, "Gary Miller was particularly interested in it."

With the jump in pressure—or "pressure spike"—the reactor containment spray system was activated. In the event of high containment pressure, as in a large loss-of-coolant accident, the containment sprays were designed to automatically come on, showering the interior of the containment with 5,000 gallons of sodium hydroxide intended to "kill" the chain reaction and fix any radioactive iodine present in the containment to prevent a dangerous leak.

But the caustic sodium hydroxide sprays also "killed" expensive machinery, so Fredrick immediately acted to turn off the automatic safety system.

"I secured the spray and was assured that the building pressure on both channels was indicating back down to where it was," Fredrick said. "The building pressure was back to normal. It could have been an instrument malfunction that caused that. It was sufficiently abnormal to be completely out of my realm of experience."

What Miller heard and Fredrick saw on his pressure recorder was a hydrogen explosion inside the containment. Some spark, possibly from cycling of the Electromatic valve at about the same time, touched off the mixture of hydrogen gas and air that had been collecting for hours in the top of the containment. The explosion would have been enough to destroy a small building, but was hardly noticed in the 600,000-cubic-foot containment with its 4-foot-thick reinforced concrete walls.

The containment walls are tested to 60 pounds pressure every three years and designed to fail at some pressure above that, yet Edson Case of the NRC admitted later that he was "a little surprised" that the containment was not damaged by the blast when the pressure spike was finally correctly interpreted two days later.

But on Wednesday it was a minor event and had no great significance to anyone in the control room. Nor did anyone realize that, paradoxically, this accidental explosion helped spare the lives of the several thousand people who might otherwise have died as a result of the accident at Three Mile Island.

7

Young Bill Scranton

Ben Livingood had plenty of time that Wednesday morning. As the Harrisburg correspondent for the Allentown *Call,* he didn't have anything to cover until Lieutenant Governor Scranton's press conference on energy policy scheduled for 10 A.M. He used that extra time to drop by his lawyer's to sign a new will. "I thought about that will more than a couple of times in the next week," Livingood said later.

Livingood had covered Harrisburg for the *Call* for six years. His beat was government and politics and he knew how to dig for a story when he had to. It would soon be time to dig.

When Livingood got to the capitol newsroom, a lot had changed. The smoky suite of rooms was all business that morning. The everpresent card game was missing. The television was turned off. Scranton's press conference had been delayed and the topic changed from energy policy to the accident at Three Mile Island. "It was the first I had heard about an accident," Livingood said, "and nobody else knew anything about it, so we just waited."

The reporters who covered the first Scranton press conference of that Wednesday were not science writers. Like Livingood, they were political specialists or general assignment reporters. None of them knew the terminology. They had never heard of a general emergency or a site emergency. Livingood said, "To those of us who cover government, we didn't know any more about Three Mile Island than to see the steam coming out of the cooling towers."

Lieutenant Governor William Scranton III took the changes of the morning in stride. He was chairman of the state Civil Defense Council as well as of the Energy Council and felt qualified to speak for the Commonwealth of Pennsylvania on the current situation at the plant. Scranton loved to speak with the press. Known as the darling of the county Republican chairmen, he would go anywhere in the state to speak to any political gathering, in sharp contrast to the reclusive governor, who could hardly ever be coaxed into such appearances.

The scion of a famous Pennsylvania political family, Scranton was following the lead of his forebears. At thirty-two, he was the state's youngest lieutenant governor and the first in modern times not to have held a previous elected office. The slim dark-haired lieutenant governor had never held a government position on any level before being elected in 1978. He had been the owner-publisher of three weekly newspapers in Pennsylvania and before that had worked in the Transcendental Meditation movement.

Transcendental Meditation didn't impress Pennsylvania voters much, but the Scranton name and the endorsement of the candidate's father—a former governor of the state and a presidential hopeful in 1964—did. After election, however, the new lieutenant governor tried to show the press that he was his own man by insisting on being referred to in all press accounts as "William Scranton III," rather than "Young Bill Scranton."

The energy press conference had been scheduled for 10 A.M., but Scranton needed more time to prepare a statement and to gather the state's experts around him. The press conference finally began at 10:55 A.M., and Jim Panyard of the Philadelphia *Bulletin,* dean of the unsympathetic capitol press corps, began the questioning by chastising the young lieutenant governor: "Do you realize that the entire press corps has been here for an hour?"

The press conference did little to enlighten the press. As Livingood remembered, "We were told in very simple terms that there had been a small release of radiation and that the situation was under control. It wasn't until later in the day that we began to get a measure of what had happened, that it wasn't just a simple little accident to be taken casually."

By 10:45 A.M., press inquiries were coming in from all over the nation. PEMA had already been contacted by reporters from the national media and by television and radio stations as far away as Portland, Oregon. Scranton felt it was time for the Commonwealth to make a statement.

Scranton's prepared statement was made in clear and confident terms: "Everything is under control. There is and was no danger to public health and safety . . . No increase in normal radiation levels has been detected."

In the questioning that followed, however, it became evident that the lieutenant governor did not know everything that was happening at the nuclear facility. Just as he was answering a question on releases of radiation that ". . . there could not have been any detected in the atmosphere," Scranton was challenged by one of his own experts.

William Dornsife, the state's nuclear engineer, facing his first press conference, spoke up in contradiction to Scranton: ". . . Before we came up here I got word that they detected a small amount of radioactive iodine in the ground . . . It may show up in the milk within a week or so like during the fallout incident we had a couple of years ago."

Scranton turned to watch Dornsife as the engineer tried to explain about the release of radioactive iodine. The explanation didn't work very well. He said later, "We were getting nowhere . . . it was obvious that my discussion of what had happened was completely over their heads."

When the press conference ended at 11:30 A.M., Scranton went immediately to brief the governor. Leaving the second-floor media center, Scranton walked down one of the ornate stairways in the executive wing of the state capitol building. On the steps, a reporter asked him to comment on a Met Ed statement that there was no off-site release of radiation. "This was in direct contradiction to what I had just learned from Dornsife," Scranton said later. "I told them to my knowledge there *was* off-site radiation. I was sticking with that story." Communication problems with Met Ed were only just beginning.

That morning Blain Fabian, Met Ed's head of communication services, had told both Chip Miller of ABC radio and Wally Hudson of the Reading *Eagle* that some radiation had been released into the environment. But, calling back later, neither reporter could get this information confirmed by Met Ed. Howard Seldomridge told Miller that radiation was confined to the containment building, and Judy Botvin said much the same thing to Hudson.

Ben Livingood had a bad feeling about the developments of the morning. "I just felt so damn stupid," he said. "I didn't understand a damn thing that was happening, and I didn't think I should trust what Met Ed was saying without knowing what all the words meant. I wouldn't say that they had been 'shady' in their dealings with the press up to that time, but they sure weren't forthright either."

After the press conference, while most of the capitol press corps ate lunch or played cards in the newsroom or went up to Middletown to interview citizens, Livingood took a long-distance lesson in nuclear engineering. Gary Sanborn, another *Call* reporter, telephoned Livingood from Allentown to get an idea of what was going on. Sanborn had been a reactor technician in the Navy aboard the USS *Enterprise* and had a personal interest in the accident. After Livingood filled Sanborn in on the situation as best he could, Sanborn answered questions and explained reactor systems to Livingood. They were on the phone much of the day.

But Livingood knew he still wasn't ready to tackle the hard technical part of the story. He asked Sanborn to cover the technical side. "We knew all the brains of Met Ed were in Reading, so Gary went up there while I stayed to cover the political front in Harrisburg."

As the Scranton press conference was ending in Harrisburg, the situation continued to deteriorate at Three Mile Island. Pressure was building up in the auxiliary building where radioactive coolant was puddled on the floor. The still-warm coolant filled the building with mist. The auxiliary building, unlike the containment, was not designed to withstand very much pressure. As the atmospheric pressure in the building increased, vents in the roof would automatically open, each time allowing a puff of steaming air to rise over the plant. The radioactive cloud would pause in the near-windless conditions, quite visible, and then slowly dissipate. It was a rare opportunity to "see" radioactivity.

Across the river at the observation center, Jack Herbein was having his first meeting with the press. By late morning, more than a hundred reporters were on the scene. Rental cars and television vans jammed the visitor center parking lot. The NBC helicopter squatted in a wheat field nearby.

Like almost all of the technical staff at Three Mile Island, Herbein had a Navy background. In fact, he had been on Navy Reserve duty in Philadelphia that Wednesday morning. Now the utility's vice president for Electrical Generation, Herbein had been the station manager at TMI and knew the plant as well as anybody in Met Ed. He also fancied himself a veteran public spokesman, having been often interviewed by the local media.

But this Wednesday afternoon, Herbein stood on a wooden platform, his back to the cooling towers of the nuclear plant, facing half a hundred reporters from the *national* media. Was he the right person to be explaining what was going on? Robert Bernero, an NRC technical briefer, didn't think so:

"I have a high regard for him as a station superintendent. He's conscientious, honest, straight, hardworking, but mean. He's a tough guy. He's the last man in the world—knowing him personally, having dealt with him professionally—he is the last man in that organization I would ever pick to be a public spokesman . . . Jack has a very tough, condescending attitude. He's a very sharp man himself. And he's the sort you would want to avoid, a very strong-willed, assertive expert who has little patience with ignorance or with slow understanding. His attitude is: 'I'm going to give you twenty-five words to explain this problem and if you don't understand it with those twenty-five words, out you go, you don't get another word.'"

The crowd of reporters shouted at Herbein, each trying to ask his question. The utility executive's tone was brusque and confident as he spent almost an hour describing the accident and answering questions. While there had been some failed fuel, "only a few pellets may have melted through" and such failure was limited to "considerably less than one percent of the fuel." Herbein said the plant would be in cold shutdown in another day and back in commercial operation within a few days. He was wrong.

While Herbein was fielding questions in front of the plant, the vents on the auxiliary building slammed open, releasing a puff of moist air into the sky. As it hovered over TMI, some of the reporters looked up, startled. "What was that?" a reporter asked.

"Nothing. Don't worry about it," Jack Herbein, the nuclear engineer, replied, turning to the next questioner.

8

A Train-Ride in the Country

Twenty-five years before, Betty Garber and some of her friends had taken a train-ride into the countryside outside of Pittsburgh. It was 1953 and Betty Garber was one of the several hundred picnickers who traveled that day to see ground broken for the nation's first nuclear power plant at Shippingport, Pennsylvania, on the Ohio River.

"It took a couple of hours by train just to get out there," she remembered. "It was great fun. We took lunches to eat in these great green fields that were going to become the power plant. And President Eisenhower pushed a button back at the White House to operate the machine digging the first hole. It was a good day. Who'd have known it would come to this?"

By March 1979 Betty Garber was across the state and far from the green fields and potato salad. She was a housewife and part-time office worker in Harrisburg. The power station she'd helped to celebrate was about to close down, its useful life over.

"That plant was at least fifty miles out of the city," she recalled. "I can almost see Three Mile Island from my kitchen window. You would think they'd have learned something in twenty-five years."

In twenty-five years, the nuclear power industry had learned to build bigger and more efficient power plants—and had learned to build them closer and closer to the populations they served. TMI Unit 2 was six times the capacity of Shippingport and only ten miles from Harrisburg.

But TMI-2 was the darling of Metropolitan Edison and its parent company, General Public Utilities. The plant was jointly owned by the three GPU subsidiaries, Met Ed (50 percent), Jersey Central Power & Light (25 percent) and the Pennsylvania Electric Co. (25 percent). It was a good example of the finest nuclear technology available in 1967. So long, involved and unwieldy is the process of licensing a nuclear power plant that adaption of later technical improvements is prohibitively expensive in time and in money: it starts the licensing process all over again, taking into account the new type of controls, safety systems, steam generators or whatever may have been improved in the decade after the plant design is finalized and before it is put on-line, commercially producing electricity.

When plans were completed in the early 1960s for the first reactor on Three Mile Island—Unit 1—there was no expectation that there would eventually be two reactors at the facility. To the planners at General Public Utilities, a single reactor on the Pennsylvania site would be plenty to serve expected electricity needs of the 1970s. The holding company planned to build an additional reactor, but in New Jersey, not Pennsylvania. Population and industrial growth showed the need for a second unit at Jersey Central Power & Light's Oyster Creek plant on the Atlantic coast. The planned Oyster Creek Unit 2 was to use a Babcock & Wilcox reactor, with the rest of the plant designed by the engineering firm of Burns & Roe.

By the late 1960s, though, labor problems in New Jersey and expected difficulties in licensing the second unit for Oyster Creek led GPU to change its plans, deciding instead to build a second reactor at TMI. The Oyster Creek design was little changed for use in Pennsylvania, though it would now be cooled by fresh, rather than salt, water. The design was not "optimum" for the new location, but changing it would be expensive, so GPU decided to make do.

In accepting the design for a second unit at TMI, Met Ed engineers had some concern about the Burns & Roe design for the plant. While both units at TMI would have Babcock & Wilcox reactors, a reactor is only part of a nuclear power plant. Architect-engineers like Burns & Roe design all of the plumbing, cooling, safety and control systems, amounting to more than half of the plant. Gilbert Associates, not Burns & Roe, had been the architect-engineer on Unit 1, so the two reactors would have little in common. And some Met Ed executives thought that Gilbert Associates was the better architect-engineer. With economics pressing and cost estimates already starting to soar, though, Met Ed accepted the building of a salt-water-cooled nuclear power plant on the fresh-water Susquehanna, designed by a firm the utility did not have total confidence in.

Unit 2's design was finalized in 1967. When the first spade of soil was turned for the Island's second reactor in 1972, it was expected that TMI-2 would be pumping electricity into the power grid at a final cost of $180 million. When Unit 2 finally went on-line December 30, 1978, the construction bill to GPU had reached nearly $800 million. This kind of inflation, in time and money, is not unusual in the nuclear industry, and because of it, the electricity that was going to be "too cheap to meter" was metered very closely indeed.

The reactor of Unit 2 was a Babcock & Wilcox model 177, a design also used at five other commercial power plants in the U.S.A. The B&W reactor is a pressurized water reactor (PWR) like almost all U.S. commercial reactors. It competes in the marketplace with other PWRs built by Combustion Engineering, Inc., and Westinghouse. Each manufacturer made sales claims for its particular unit, and the B&W sales incentive was economy of construction and operation. The smaller Once Through Steam Generators featured in B&W reactors were cheaper to build and allowed higher steam temperatures, increasing system efficiency. They also boiled dry much more quickly than other steam generator designs and required much more operator attention for this reason. Both of the Once Through Steam Generators in Unit 2 boiled dry within the first two minutes of the accident.

Higher steam temperatures allow greater thermal efficiency in a steam plant and this, along with other design features, allowed the model 177 to save 1.5 percent in nuclear fuel costs over other designs —1.5 percent that could be translated into millions of dollars over the lifetime of the reactor.

Another time and money saver was the different way that B&W plants were controlled during a "transient," or significant unexpected change in reactor operation. In most other reactor designs, a turbine trip automatically activates safety systems to trip or scram the reactor. This shuts down the power plant for the several hours it takes for certain fission products to decay enough for the chain reaction to be re-started. In the B&W reactor, turbine and reactor controls are not connected, and the reactor trips in response to high system pressure, thus providing a 3- to 4-second delay before the reactor scrams. This delay is often enough to "ride out" shorter transients and avoid costly reactor downtime. Of course, it also puts great importance on the Electromatic safety valve that has to relieve high system pressure. And Electromatic valves have been known to fail.

Unit 2 "went critical"—began the nuclear chain reaction—on March 28, 1978. In the several months that followed, B&W and Met Ed engineers worked the bugs out of the new reactor, testing reactor systems and building up to steady full-power operation before the unit went on-line and was committed to providing electricity to the power grid. Much of this breaking-in period was spent with the reactor out of action as subsystems were checked and repaired. Before going on-line December 30, 1978, Unit 2 had been down for repairs 195 of the 274 days since it went critical—71 percent of the time. The industry average for downtime before going into service is 40 percent.

"Nobody ever did like Unit 2," said a technician who still works at TMI. "It just always was a disaster. The equipment never ran right."

On April 23, 1978, a month after going critical, Unit 2's main steam relief valves failed. The resulting reactor scram was so severe that operators feared the reactor core had been uncovered. Repair of the $54,000 valves took until September. On September 20 and September 25, failure of feedwater valves caused two further reactor scrams with little damage.

On October 13, 1978, failure of the Electromatic valve on the pressurizer scrammed the reactor. This was followed by two more scrams in November and four in December—three of them in a single day.

73

Dave Labby quit his job as a first-class repairman at TMI "because I was dissatisfied with management and safety precautions and short-cuts."

Richard Blakeman resigned his job as one of Unit 2's mechanical maintenance foremen because of poor maintenance procedures and mandatory overtime work. "I've seen guys work around the clock, 24 or 32 hours," Blakeman said. "You just go, go, go. It's crazy."

Mandatory overtime ran as high as 1,200 hours a year per worker. "The power industry," said a Unit 2 foreman, "believes that it's cheaper to work its people long hours than to hire more people." There are no regulations limiting the number of hours per day or year workers can be on the job in the nuclear power industry.

As costs mounted on Unit 2, GPU came under tremendous financial pressure to get the unit on-line and making money. The accounting firm of Touche Ross, Inc., reported on GPU finances to the New Jersey public advocate:

"As a result of underestimating the scope of nuclear construction projects, GPU was unable to support the overall generation-construction program with adequate financial resources."

Rather than cut stockholders' dividends, a prospect termed "disastrous" by company officials, GPU decided to rush completion of Unit 2 while simultaneously cutting funds for the completion by 22 percent. "It was a rush job," one former mechanical maintenance worker who was a union official said, "They were on you all the time [saying] 'Let's get it done; we got to get back up there; we got to get on-line.'" In December an NRC inspection team found thirteen areas of "improperly or inadequately completed operating procedures," including problems in the reactor cooling system and the core-flood system.

As 1978 came to an end, there was some incentive at Met Ed to get Unit 2 on-line. The plant had been expensive to build and getting it on-line and into the rate base would start paying it off. Twenty million dollars in tax depreciation and $17 to $28 million in investment tax credits would be lost if the plant didn't get on-line in 1978.

Twenty-five hours before the end of 1978, Unit 2 went into service, thus qualifying the utility for the tax credits and depreciation. Some thought that the saving of at least $40 million might have had something to do with the pre–New Year rush to get the unit on-line. But Met Ed president Walter Creitz said that any problems in the plant had nothing to do with a rush for money. "We certainly never would have put it in service unless we were convinced it could be safely operated," Creitz said.

By putting it into service, Met Ed was claiming that Unit 2 would hold its own and reliably produce electricity at its full rated power. Yet in the 87 days between going on-line and March 28, 1979, Unit 2 only ran the equivalent of 40 full-power days. In January the reactor was shut down for two weeks with problems in the cooling system. Even on the day of the accident, it was not able to run at full power.

By 1979 there were 72 operating commercial nuclear power plants in the United States. The industry prided itself on a spotless accident record. Its advocates in trade groups and lobbying organizations like the Edison Electric Institute and the Atomic Industrial Forum liked to point to barrels of oil saved, air pollution avoided, and to an "accident-free" industry. But there had been a few accidents before Three Mile Island, and some of them had even included the threat of fuel melting:

December 1952. An operator error caused partial core melting and heavy contamination at the Canadian NRX research reactor at Chalk River, Ontario.

October 1957. An operator error caused core overheating and fire in a British plutonium production reactor at Windscale, England. At the height of the fire, 11 tons of uranium were burning. The accident caused contamination of an area of more than 500 square kilometers with radioactive iodine and for some days milk from dairy herds in the region was impounded. There was no evacuation.

November 1965. An operator error caused overheating and melting of some fuel in the Experimental Breeder Reactor-1 facility at the National Reactor Testing Station in Idaho. There was no radiation release or exposure.

October 1966. A partial core meltdown occurred in the Enrico Fermi-1 fast breeder reactor in Detroit. There was no release of radioactive materials, although critics of nuclear energy insist the accident came to the very edge of nuclear disaster.

January 1969. A loss-of-coolant accident in the first Swiss power reactor, located in a large cave at Lucens, resulted in loss of the reactor. There were no casualties or noticeable external contamination.

March 1975. A fire caused by careless technicians cut off many control circuits for two nuclear power reactors of the Tennessee Valley Authority at Browns Ferry Station in Alabama. The fire disrupted controls for the emergency core-cooling system of Unit 1 and caused some concern of meltdown risk. However, this did not occur and there was no radioactive contamination.

March 1977. An Electromatic relief valve stuck open following a reactor scram at the Davis-Besse nuclear power plant near Toledo, Ohio. The stuck valve was noticed by operators, and the reactor, manufactured by Babcock & Wilcox, was only slightly damaged.

Certainly the industry was not "accident free." At the time of the accident at Three Mile Island, though, there were no specific deaths that could be directly attributable to nuclear power generation in the United States.

The nuclear power industry had firm support from the majority of Americans. This source of power that had once promised to be "so cheap that it didn't need to be metered" still looked good to a people who thrilled to technology and supported any move to decrease American dependence on Middle Eastern oil without demanding changes in personal lifestyle.

And the industry worked hard to present itself in a good light to the American people. Millions of dollars from utility bills were used to support a high-class image. Hill and Knowlton, the world's largest public relations firm, supplied the words and pictures and the platoons of smiling, clean-cut men. They worked to get good press for the industry. And if good press wasn't available, no press was the next best thing.

The reactor accidents that had taken place in the United States were covered poorly by the press. The stories were not visual enough for television, and most newspapers didn't have reporters with adequate background to write about the technical details. And the industry did a good job of putting off the press during accidents. The 1977 transient at Davis-Besse was only mentioned to the public through a two-paragraph story in the Cleveland *Plain Dealer.* There were no follow-up stories.

The 1975 Browns Ferry fire produced a number of TVA press releases concentrating on the cost of replacement power and only casually mentioning that the station's emergency core-cooling system had been disabled. Most newspapers in the Tennessee Valley used the press releases without question or elaboration. The Tennessee Valley has seen its per capita income rise from 45 percent to 75 percent of the national average mainly because of the TVA, and local publishers who are, after all, businessmen, were loath to criticize their benefactor.

In contrast to the symbiotic relationship of the Tennessee Valley and TVA, Metropolitan Edison had no close relationship with the people of Harrisburg. The press saw no local interest in the company and found it difficult to get material from a utility that "didn't even have an office in town." Worst of all, however, was the public information operation of the utility that many local reporters viewed as "hopeless."

Blain Fabian, head of communication services in Reading, was saddled with producing weekly press releases about problems in the Three Mile Island plant. These releases were not required by the NRC or any other regulating body, but were put out by Met Ed at the insistence of Vice President for Electrical Generation Jack Herbein.

The weekly press releases were written by engineers rather than journalists and were almost impossible for the lay public or press to comprehend. What had been an attempt to produce information in the public interest served instead to desensitize the local press to problems at Three Mile Island. Like many local newsmen, Mike Pintek of WKBO admitted that the Met Ed weekly releases went "right in the circular file—unopened."

The press complained about the releases but they went unchanged because Blain Fabian, who had held his job for seventeen years, was of the belief that "one doesn't attempt to correct a Met Ed vice president."

Though Metropolitan Edison did not produce much material of use to the local press, there were stories about TMI in the news. Rate increases were often sought to cover the ever-rising costs of the nuclear plant, and the local press covered all the rate hearings at the state public utilities commission. There was the story of Unit 2's coming "on-line"—going into official production of commercial electricity—on the last day of 1978, just in time to qualify the company for a $40-million tax credit that year.

Worst of all, in the view of Metropolitan Edison, was the story about TMI that appeared in *Harrisburg,* a tabloid published on a shoestring by a longhaired group of part-time radical journalists. Their editorial offices were on the rundown second floor of a tenement on Harrisburg's Peffer Street. Downstairs were the two-room offices of TMI Alert, the local antinuclear group. The story that appeared in the publication's August 1978 issue was titled "Tomorrow's Disaster at Three Mile Island." It graphically detailed a hypothetical loss-of-coolant accident at Unit 2, leading to a meltdown during which:

"The reactor at TMI-2 expanded to include the outside world. A Gaussian plume released colorless, odorless, tasteless radioactive destruction into the wind. On the landscape—cows, homes, grass, trees, people . . . there was no escape. Disbelief, terror, confusion, and above all, helplessness prevailed."

Citing the Atomic Energy Commission's WASH 740 report on the outcome of a reactor meltdown, the *Harrisburg* story by Larry Arnold postulated the deaths of 27,000 people in the Harrisburg area.

The story aroused great interest among *Harrisburg*'s several thousand readers, and almost caused the magazine's demise as well. It caught the attention of Met Ed president Walter Creitz, who wrote to U.S. Congressman Gus Yatron to question the federal funding that had helped the nonprofit community publication to meet its payroll for the previous six months.

"I wasn't very warm to the article," said Creitz. "We didn't really need that. What we need is help." Creitz called the story "sensational reading, not true, a horrible article" and "distorted."

Yatron forwarded Creitz's letter on to the U.S. Department of Labor, which ordered that federal funding be stopped for the Harrisburg Independent Press, publisher of *Harrisburg,* "following a congressional inquiry." With their federal funding cut, some staff left and the rest continued without pay to be the only voice in central Pennsylvania pointing out that there were no evacuation plans and no reliable means of even monitoring radiation in the event of a real accident at TMI.

9

Harrisburg

Silver-haired, with too much hair spray and make-up, the waitress leaned over the lunch counter of one of the little restaurants near the state capitol in Harrisburg. She had been asked about the accident at Three Mile Island, but she was in no hurry to answer.

Adjusting glasses and plates on the Formica-topped counter as she had for at least thirty years, the fiftyish woman ignored the radio sitting next to the breakfast cereal display. It was tuned to WHP and Jim Moyer was telling Dauphin County about "releases of radiation to the environment." A line of retirement-aged men sitting at the counter nodded into their cups of coffee. The waitress pulled an order pad out of her pocket as Moyer's voice said something about a "cold shutdown in a few more days."

"I've lived through four bad floods in this town," the waitress finally said seriously. "Every time the water rises, it always goes back down again, and that's the way it is here. That factory or plant out there isn't gonna hurt me. My boy helped build it and he says it's okay. Are you ready to order yet?"

If Metropolitan Edison played down the danger at TMI in its public statements, it was doing what most of the people of Harrisburg wanted. The town didn't really care about nuclear power. Filling potholes in the state's highways was the major issue of the day. The main industry in this city of 60,000 was running the Commonwealth of Pennsylvania.

Harrisburg lay along the north bank of the Susquehanna River about nine miles west of Three Mile Island. On a sixty-foot knoll in the center of town, somewhat grandly called "capitol hill," stood the ornate green-domed, late Victorian statehouse opened by President Theodore Roosevelt in 1906. It dominated the city.

The capitol was surrounded by modern state office buildings filled with workers who drove in from the suburbs. The state workers' exodus from the city killed its commercial life. Dozens of empty storefronts opened on the streets of Harrisburg, creating a dead zone around the state buildings. The stores had moved to malls near the edge of town to be nearer the homes of those with disposable income.

With the stores and state workers moved to the outskirts of town, most of the inner city was inherited by a multiracial lower class. Former mansions crumbled along treelined brick streets. Lines of frame Victorian row houses leaned against each other in general neglect. Landlords lived elsewhere, and the tenants, often young and always poor, couldn't afford to repair their homes.

In contrast to the decaying neighborhoods just west of the capitol, a line of mansions faced the Susquehanna, looking across the river to Camp Hill, where many of the commuters lived. The Governor's Mansion was among those facing the river. Built in 1968, the large building was made of tan-colored brick and stood behind a black iron fence. Two blocks away, abandoned houses sat, covered with graffiti, their broken windows airing empty rooms.

By midafternoon on Wednesday, the Harrisburg *Evening News* had hit the streets. Mary Bradley's first story about TMI added a print message to the story that had dominated local radio and television all day.

Echoing the confidence of the utility spokesmen, the story looked back at the accident as though it were over. With little trouble, Met Ed was expected to have the plant shut down, "in a couple of days." This is what most of the *Evening News* readers had expected. They'd lived at peace with TMI for years.

Mrs. Everett Quackenbush, whose home was within sight of the plant, said that afternoon, "The ecologists will play this up, but the plant has never hurt us and I don't think it ever will. I've heard how thick those walls are."

One woman whose husband was among the five hundred local people employed at the plant said, "I always hated to think what might happen . . . so I didn't." Her husband was bald and had had a vasectomy. He belittled the danger, saying, "I've already lost all my hair and I'm sterile. What else can they do to me?"

Holly Garnish, who lived directly across the river from the plant and heard the first blast of steam released from the turbine that morning, was resigned to nuclear power and to TMI: "I have mixed emotions about the plant . . . I'm angry when something happens like today, but I've learned to live with it.

"Really, it was very depressed around this area until they started building Three Mile Island. It's meant so many jobs for so many people that I can't help but think that it's been good for the area."

Rita Connor, who lived a half-mile from TMI, had already assured her mother and sister in New York that she "hadn't been blown to bits." "I don't worry about it," she continued. "You see these lights burning in this house?" she said, gesturing around her. "As long as I want to keep them burning there have to be plants like this some-place."

In Royalton Township, Monica Drayer shared Rita Connor's fatalistic support of nuclear power. "If you are going to go, you're going to go. Life is one big gamble . . . That plant's as safe as anything else. I really don't think the government would let them do anything to endanger our health."

But not all the people around Three Mile Island were sure that the trouble was over. Arlene Marshall, an office manager in Harrisburg, said, "We've heard nothing. We've heard that it is safe. We—we're all going to be okay and we have nothing to worry about, and I don't think anybody really believes that.

"I never went out and carried signs and protested nuclear power plants. I think that I am against them. I think that I am afraid of them. At the same time, I don't want to give up my dishwasher and my dryer and my air conditioner, so I don't know the answer."

And an elderly lady working behind the counter at the general store in nearby Falmouth was still very worried. "I don't like it," she said. "I don't like it at all . . . It's almost enough to make me want to sell out and move."

Tom Gerusky, head of Pennsylvania's Bureau of Radiological Protection, sat holding a Geiger counter in his office in the two-story brick building that housed the Department of Environmental Resources, of which his bureau was a part.

"I keep a Geiger counter behind my desk," Gerusky said. "This time I pulled it out and took a reading right there in my chair. It didn't read much, but it was slightly more than I would have expected background radiation to read.

"I took the counter down to my car and started the drive to Three Mile Island. Our office is on the far side of Harrisburg, sixteen miles from the power plant, and as I drove through the city, I saw the radiation level slowly start to rise. That pretty much told me that the level I was reading came from the reactor.

"Driving east, the meter rose until it stopped at three or four millirems. I looked around and found myself parked right next to the plant."

After his early afternoon meeting with the press at the observation center, Jack Herbein drove into Harrisburg with Gary Miller to brief the lieutenant governor. His hour of questions and answers had been difficult. And the motley staff of *Harrisburg* was at the observation center, distributing copies of their TMI meltdown story to the press.

Distrust of Met Ed was common in the governor's office by mid-afternoon. When Herbein and Miller came to brief Scranton, Paul Critchlow, Governor Thornburgh's press secretary and close adviser, asked for an assistant state attorney general to be present. Also there were Critchlow, Tom Gerusky, Colonel Henderson of PEMA, and Mark Knouse, Scranton's chief aide.

Herbein's first-ever meeting with the lieutenant governor was less than cordial. Scranton was angry. He'd given the press information from Met Ed at his first press conference, saying that there had been no off-site release of radiation, only to be contradicted by one of his own state agencies. He wanted Herbein to explain. "Have you been venting radiation?" Scranton asked.

"Herbein came up here and told us that they had not lied to anybody," Scranton said later, "that there was venting . . . He said he didn't know of anybody from Met Ed that had said there was no off-site radiation."

The Met Ed vice president squirmed and tried to dodge the sharp edges of Scranton's questions. He and Miller had "authorized" the venting of gases from the auxiliary building. Actually, there was no way they could have stopped the venting, since the building was not designed to hold gases under pressure. But Herbein didn't want to admit that he and his company were not in complete control at TMI.

Herbein's purpose in coming to the meeting was soon obvious to those there. "Herbein was trying to calm everyone and make sure that he could restore, at that point, a rather tarnished faith that we had in Met Ed," Mark Knouse recalled.

Scranton asked Herbein why he hadn't told the press about the radiation releases while he was briefing them out at the observation center that afternoon. After all, he'd spoken to reporters for an hour in clear sight of the stricken plant. Herbein said that he hadn't told the press about the releases "because they hadn't asked."

Scranton "realized then that the problem was getting pretty large and that we couldn't count on anybody at Met Ed for any type of information. And we wanted to get hold of some NRC people to find out exactly what was happening and we wanted some backup monitoring system."

Looking back at that meeting, Critchlow recalled: "I knew then that their M.O. was likely to be minimal—was likely to minimize all these kinds of events. Right from that moment on, we virtually had nothing to do with Met Ed."

Toward the end of the meeting, someone suggested that Scranton and Herbein make a joint statement to the press, but this was vetoed by Critchlow, who said, "I have deep suspicions about Herbein and I am not going to associate the lieutenant governor with him." The state and utility never did make any statements together.

Tom Gerusky, of the Bureau of Radiological Protection, was present at the meeting between Scranton and Herbein. Gerusky knew there were continuing releases of radioactivity from TMI. And he knew that the releases were not controllable by Met Ed.

After his meeting with Herbein and after what he had learned from Gerusky about the continuing releases of radiation from the plant, Scranton decided to go back to the capitol media center and meet with the press for the second time that day.

Scranton was angry and wanted the reporters to know that. He was not going to make an attempt to minimize the danger of the accident in this second press conference at 4:30 P.M.

"This situation is more complex than the company first led us to believe," Scranton's opening statement read, in part. "We are taking more tests. And, at this point, we believe there is still no danger to public health.

"Metropolitan Edison has given you and us conflicting information . . . There has been a release of radioactivity into the environment. The magnitude of the release is still being determined, but there is no evidence yet that it has resulted in the presence of dangerous levels . . . The Pennsylvania Department of Environmental Resources was not notified about the release until about the time that it was halted . . ."

The questions from the press that followed were more thoughtful than they had been at Scranton's earlier appearance. The accident seemed more serious, and some reporters had taken time to learn more about what they were covering.

One of these was Ben Livingood. His questions were pointed and informed and startled even the other reporters. "The big question, the one I needed to ask," Livingood said later, "was 'Was there any evidence of fuel damage?' I think it surprised Scranton and he didn't know what to answer at first. And I wasn't any better prepared than he was, because I wasn't sure what it meant either."

Livingood's rapid exchange of questions and answers came with William Dornsife, the state's nuclear engineer, who was there along with Scranton, Gerusky and PEMA's Oran Henderson to answer questions.

LIVINGOOD: I have a couple of questions. Number one, has the company given you any indication that they found any evidence of fuel-element rupture?

DORNSIFE: Yes. They have said there has been evidence of—I'm not sure if it's technically fuel-element rupture, but some element of fuel decay —cladding damage—which is something they say is under control. It's a zirconium metal that encapsulates the fuel.

LIVINGOOD: So, it is a breakdown in the encasement rods, right? So, radioactive particles that were encased in those rods are escaping into the primary coolant. The primary coolant, somehow or other, is getting into the secondary system. Isn't that true? The presence of iodine. Where did you detect the iodine this morning? Was that at Goldsboro?

DORNSIFE: Yes.

LIVINGOOD: That would suggest, would it not, that there was a release of primary coolant, right?

DORNSIFE: Yes. To get any [radio]activity out, it had to be primary coolant.

LIVINGOOD: How did that primary coolant get out of the primary containment building and into the secondary cooling system—which is where the steam is vented, right? How did that secondary system get contaminated?

DORNSIFE: According to the company—And I think that in these kinds of questions you are going to ask the company because we didn't have somebody there.

ANOTHER REPORTER: The company won't answer the questions.

When Livingood reported to Sanborn that there had been some fuel damage, he was told that this meant some of the fuel had melted. "Gary said then that it could only get worse and that the thing could still melt down right as we argued about it in Scranton's office."

10
The Media

Late afternoon on Wednesday saw more than two hundred reporters from every medium at the observation center across from the plant. Bill Gross was enjoying his moment in the sun. "I read the statement we'd written that morning," he said. "Then they wanted me to read it again for the cameras, another time for the radio, finally with my back to the plant."

Several television camera crews were based at the observation center, with more on the way. Even public television decided to base the MacNeil/Lehrer Report at WITF in nearby Hershey with that night's opening done direct from the observation center.

For all the activity across from the plant, there wasn't very much news being released. Met Ed was saying little and the employees waiting at the observation center were often reluctant to talk to the press—already they were being discouraged from speaking by Met Ed. So the newsmen interviewed the people who lived near the plant and even interviewed each other.

Mike Connor didn't go to grade school that day. He wanted to open a hot-dog stand for the reporters and utility workers, but his mother said no.

John Garnish claimed he'd been interviewed by ABC and five local television stations, by the Washington *Post, Newsweek,* and by the New York *Daily News.* *"Sixty Minutes"* will be here any minute," he said, obviously enjoying the attention.

Reporters need telephones to do their work, and the observation center had only two—one of them a pay phone that stopped working when its coin box filled about midmorning. Two hundred journalists tried to file their stories from an out-of-order pay phone.

Among the television contingents, NBC probably had the largest staff on-site, but Ward of CBS had sent in a second crew at midafternoon and had another crew with correspondent Robert Schakne trying to get information from the NRC in Bethesda and Washington. "We couldn't get anything from the NRC in Harrisburg and I already distrusted the utility," Ward said, "so Bethesda was our only real source."

Ward knew the accident was important when off-site radiation releases were finally confirmed by Met Ed. What bothered him especially was talk of "building shine"—gamma radiation beaming through the 8-inch-steel walls of the reactor vessel and 4-foot-thick walls of the containment building. "It takes a very hot containment to shine through all that concrete and steel," Ward said. "When I first heard them talking about 'building shine,' it really put my teeth on edge."

A sort of "radiation macho" had already developed among some of the reporters in Harrisburg. The he-men made jokes, but some of the reporters were already getting scared.

"I'd done stories on low-level radiation," Ward said later, "and briefed my crews going out that there is no such thing as an insignificant level of exposure. We have a lot of young women camera persons, and I told them not to go if they thought they might be pregnant or planned to have a pregnancy soon."

Despite Ward's concern about low-level radiation, to many reporters the story seemed almost over. And this feeling may have been founded on the confidence of Met Ed's Jack Herbein who would not admit that the accident was bad or that the operators had been at fault:

HERBEIN: I'm satisfied that all my employees acted properly.
REPORTER: Would you characterize this accident as a close call?
HERBEIN: No. I wouldn't characterize it as a close call . . . Things are falling off right now, as I've indicated. The coolant-injection systems

are functioning properly. We've seen the plant pressure coming down and we expect soon to be on our low-pressure decay-heat-injection system. At that point, the plant will be in a cold shutdown condition.

Much of the editorial staff of *Harrisburg* had spent the day at the observation center. As long as they lasted, copies of the TMI melt-down article were distributed to the press and a few interviews were given. "I don't think that Met Ed liked having us there," one *Harrisburg* staffer said later, "but there wasn't a hell of a lot they could do to keep us away."

The press needed places to work, so Saul Kohler, who'd started his new job as editor of both the Harrisburg *Patriot* and *Evening News* just that morning, threw open his newsroom to any reporters who needed space and a phone. The 80-foot-square newsroom was crowded enough with two daily papers running from it, but soon 60 to 80 additional reporters filled the room, with its dirty gray and cream enamel walls and scarred linoleum floors.

The television situation was no better. WHP, the local CBS affiliate, made room for the CBS network crews under Ward, but Harrisburg had no NBC affiliate, so both NBC and ABC worked out of the small studios of WTPA, just down the street from WHP.

The network newscasts that evening generally shared the optimism of Met Ed. They looked back at the accident as if it were over and asked important questions about what this event meant for the future of nuclear power. But Walter Cronkite even made retrospect seem exciting:

"It was the first step in a nuclear nightmare as far as we know at this hour, no worse than that. But a government official said that a breakdown in an atomic power plant in Pennsylvania today is probably the worst nuclear accident to date . . ."

One of the issues to dominate media interest in the late afternoon, when the problem of public health seemed to have subsided, was whether any of the plant workers had been contaminated with radiation. According to ABC:

"Jack Herbein, a Metropolitan Edison vice president, refused to say how serious the contamination of the plant—south of Harrisburg, Pennsylvania—might be, but claimed 'It's nothing we can't take care of.'

"Outside the plant, on the Susquehanna River, ABC's Bettina Gregory talked with John Currick, an employee of a subcontractor doing work at Three Mile Island. Currick claims power company workers at the plant are not as careful as they should be about radiation."

> JOHN CURRICK: It's really scary, and at times it gets pretty well crapped up in there and you can't get in, and it takes a while for it to get cleaned up.
> BETTINA GREGORY: Wait a minute. When you say it gets messed up in there, what do you mean?
> CURRICK: Contaminated. Contaminated.

With the networks already looking at the accident in retrospect, statements of outside experts became increasingly important. National Public Radio's *All Things Considered* news program used an antinuclear spokesman to point up the confusion within the NRC. Robert Pollard, a former licensing project manager with the NRC, quit his job with the federal agency in a dispute over the safety of commercial nuclear plants. It was Pollard's contention that the NRC had licensed many unsafe nuclear plants under political pressure. *All Things Considered* asked if Three Mile Island's Unit 2 was one of those unsafe plants.

> ROBERT POLLARD: Yes. Because you have to understand the problems that the Nuclear Regulatory Commission has identified and reported to Congress apply to all plants by the same manufacturer.
> Take, for example, this Babcock & Wilcox design, which is Three Mile Island. One of the safety problems the NRC has identified is what's called interaction between systems. The Babcock & Wilcox design, like the Westinghouse design, has a lot of interaction. The end result is that a single failure or a single malfunction can both cause an accident and disable the protection against the accident.

The NRC, already coming under fire from the press, was less than decisive in its response to Pollard's contention: "Well, that is his opinion and he is entitled to his opinion, but the Commission itself thought the plant was safe when they licensed it."

For most of the afternoon, the reactor of Unit 2 was without any cooling system at all. The main coolant pumps—four of them at 9,000 horsepower each—would have pumped only steam if they had been started. And steam might have been enough. It probably would have helped to cool the core if steam had been circulated and might have even helped to eliminate the steam bubbles that prevented natural circulation from cooling the unit.

But the main coolant pumps were not on. There was some doubt at Met Ed whether they could be turned on at all. There were problems with the oiling system on the pumps and a few trial starts had not been considered successful. Starting the pumps without an oiling system working could result in seal failure and coolant leaks through the pumps themselves.

Basically, though, the pumps were off because the operators had been taught that the system could cool satisfactorily without them. And the company would frown on damage to their multimillion-dollar pumps. They would be needed when the plant went back on-line in a few weeks.

Throughout the morning and early afternoon, those in the control room had to use respirators off and on as the wind changed direction and blew clouds of radioactive xenon gas through the building. Activated charcoal filters were installed to take any dangerous radioactive iodine out the the air supply, but the xenon swept right through the charcoal. And so would have the iodine, if there had been any around, because the activated charcoal filters that protected the air supply of the control room were only operating at about 25-percent effectiveness. Not only were they filled with the wrong grade of activated charcoal, they were worn out. The filters were designed to swing into operation only when high radiation levels were detected, but had accidentally been going twenty-four hours a day for more than a year. The activated charcoal filters were almost useless.

91

But radiation levels in the control room made little difference to Gary Miller. The Three Mile Island station manager and emergency director was determined to stay on under any circumstances and cure the problems of the Unit 2 reactor.

"You cannot get me in a situation where I would leave," Miller said later. "You could not walk away from this thing and say it is going to shut itself down . . . I could have stayed up even if the reading had gotten pretty high. I would have stayed up there a long time and switched crews to minimize health risk . . . I would never leave the control room from a radiation standpoint."

It was for other than radiological reasons, then, that Gary Miller left his reactor that afternoon to go into Harrisburg with Jack Herbein and meet Lieutenant Governor Scranton. During the two and a half hours he was gone, the reactor continued to sit without any cooling.

Not all the action at Three Mile Island was taking place in the control room or at the observation center. Many important jobs had to be performed elsewhere in the plant—some of them made dangerous by high radiation levels.

Throughout the afternoon and evening of March 28, auxiliary operators and other workers at TMI regularly entered areas of high radiation. Sometimes they took adequate precautions before entering: wearing full protective clothing and breathing systems, carrying radiation meters, wearing extremity dosimeters to measure how much radiation is received by hands that pick up coolant samples or by feet that walk through radioactive puddles. More often, though, one or more of these important safety elements was missing.

An auxiliary operator was told to recharge the core flood tanks with nitrogen, normally a 5- to 10-minute job. He wore protective clothing and a breathing system, but didn't have a required high-range pocket dosimeter.

Entering the auxiliary building, he climbed to the second level, where the nitrogen valve was. Radiation levels varied as he walked through the building—20 rems per hour, 60, 100, 30 and, finally, 10 rems per hour at the valve.

Back outside after completing the job, it was obvious that the man was heavily contaminated with radioactive particles on his clothing and skin. The usual decontamination procedure was a quick shower, but the auxiliary operator decided not to bother getting decontaminated twice and simply went back into the auxiliary building to attempt a second job—a transfer of liquid from the rad-waste tank.

Reentering the building, he went through the switch gear room, which was hopefully a bit less contaminated than alternate paths to the rad-waste panel. The pump wouldn't start at first, so he waited three minutes and then tried again. One more time and the pump still wouldn't start, so the auxiliary operator gave up his dangerous attempt.

He showered in the plant's decontamination area, wearing a mask to avoid inhaling any of the radioactive particles washed off his skin. Back in the Unit 2 control room, a radiation meter called a "frisker" went off as he passed the desk it sat on, even though it was after the decontamination shower. The man had received 3.170 rems in his total 15 minutes of work. He was "burned out for the quarter," having exceeded his allowable radiation dose of 3 rems for a three-month period.

Burning out wasn't unusual that day. Auxiliary operator Ron Fountain watched it happen while he waited his turn. "A couple of fellows went out and came back a few minutes later," Fountain said. "They were burned out quick. We had to go out alone, not in teams the way we usually do it."

Fountain's job was to open an auxiliary spray valve. To do this, he had to climb a handrail up to a pipe run where the valve was located, 47 feet above the floor. The simple job quickly turned into a terrifying experience as Fountain tried to fight panic while getting his job done in the tasteless, odorless radiation field:

"I had to get up to the second level to open my valve but the elevator was broken," Fountain said. "I didn't want to run out of air or time and I didn't have my watch.

"I knew where I was going and I knew which valve I had to open. It was a locked valve. I had no intention of entering any rooms; so I thought it'd be—I had no idea at the time what the radiation levels were. As I got to the aux building, an HP [health physics] fellow came out. He told me the radiation levels in the hallways were 100 rems per hour and I asked him at that time for his Teletector [radiation detector]. But he still needed it and could not surrender it. So I just told myself: time, distance and shielding. In other words, bag it through the hallways, and once I got in the area of the valve . . . it was pretty well shielded. I thought I'd be all right.

"Then the game was getting Scott Airpaks, so I used the [oxygen] bottle that was half full. There was no one there to assist us; no one there to help you. I climbed up to the valve.

"I was afraid. I should have walked, but I started running. I started hyperventilating. My instinct was to rip off my mask but I knew the air was heavy with particles. I said a prayer. I had to gather my wits. I was sweating and breathing heavy. I made my walk to the valve. I opened it.

"Then I walked toward the door. The Airpak's two-minute warning was ringing. Then I ran out of air, so I had to remove the mask just, oh, maybe twenty-five to thirty feet before the door where you get out. I was bursting through the door, kicking it open. I ripped the mask off, put on another. Usually we've got other personnel to help us. But not that day.

"So I got out of there and got undressed. And everything was contaminated. There was no clean area you could use. So I went upstairs and as soon as I opened the door the monitor they had in there . . . the frisker in there went off. So I told them I had the valve open and left."

In five minutes, Ron Fountain had been exposed to two rems of radiation.

A day later, Fountain explained what it was like to be an auxiliary operator at TMI:

"Most of us are having trouble sleeping. We've all been running on adrenalin. You come home, go to bed, toss and turn, can't fall asleep, pace the floor. I'm concerned about what's going to happen to our jobs. We don't know yet what's happening in the reactor. It's still not under control. There are so many things going on, nobody knows the full story yet."

*   *   *

With a direct phone line finally established to the Unit 2 control room, the NRC incident response center in Bethesda was at last in action that afternoon. The two major rooms of the center were filled with the technical brains of the NRC, connected by 26 telephone lines to other experts around the country. As information came into the main room of the IRC, it would be relayed into the glass-walled conference room that housed the Executive Management Team—the men authorized to commit the resources of the NRC and to speak for it officially. This hand-relay of information from room to room had its problems, as Edson Case, a member of the Executive Management Team, related:

"The system is supposed to be that they get all that information over there, and then periodically like—depending on the severity of the thing—more like every fifteen minutes, or if it's quiet, every half-hour, some one guy would gather all the information that they'd gotten in that period of time and come in and brief us.

"I always thought it was a dumb system. In the first place, there's a lag. In the second place, it's filtered, and what you've really got to do in deciding what to do, if anything, is to take into account the reliability of the information that you receive. And if it's filtered through somebody, you lose that touch. If the information comes in from some guy that I have confidence in, I am more likely to believe him than some other guy.

"It's just too goddamn organized. If you want to ask a question, you're supposed to write it down on a piece of paper and pass it to the coordinator, who gets the answer. I could never follow the system, so I just went in there and asked."

The NRC was only just learning to mistrust Met Ed's evaluation, and so Case passed on with only a small note of caution what was generally the company line when he spoke to reporters that afternoon:

"We were expecting them to go on decaying, cooling, residual heat removal. And they weren't—and that should have been relatively easy to accomplish and yet bring temperature and pressure down. And it wasn't—the temperature wasn't going down as fast as expected, and there was some speculation that it—the reason that it wasn't going down, that there was some sort of steam bubble."

Case mentioned nothing about operator error having been involved in the accident, nor did he point out that the high-pressure injection system had been cut off. The fact that cooling-water injection had been stopped didn't seem so important to Case at that moment, but it was very important to Vic Stello, who was in the room with him.

Stello was head of the NRC's office of Inspection and Enforcement. It was his job to make sure that 72 operating commercial reactors in the United States followed the book. His inspectors observed in control rooms, inspected valves, pumps, gauges and pipes or tried to sneak past security systems and simulate sabotage.

And Vic Stello was good at his job. He understood the workings of nuclear reactors very well—probably much better than did most of the inspectors under him. That was the problem with the office of Inspection and Enforcement, or I&E. The inspectors did not have time to get to know each system they were supposed to inspect. Standing in the control room, they could often not tell if the operators were performing the correct operation for a particular reactor problem—they could just tell if the operators were performing the operation specified in the book. The inspectors were not licensed operators themselves. They were bound to a strict set of Emergency Procedures. The operators had just to follow the specific Emergency Procedures in the book, and the inspectors were there to ensure that the Emergency Procedures were followed, whether they actually cured the problem or not.

Stello was a volatile Italian and prone to shouting. He began to shout that afternoon soon after learning that the high-pressure injection had been turned off.

"Through most of the afternoon," Stello said, "we were trying to advise the licensee that he may have a condition of inadequate core cooling and that there would be a need to get more water into the core to cool it."

At one point in the afternoon, Stello reached someone in the Unit 1 control room and told him he thought the core was uncovered and that the flow rate in the high-pressure injection pumps should be increased from its present level of 130 gallons per minute to somewhere near its maximum of 1,000 gallons per minute.

NRC's Richard Vollmer remembered later, "I certainly know from people who were there that Stello was yelling into the phone that the operator should understand clearly that the core was uncovered."

But Stello was alone in the Incident Response Center in thinking that the reactor core was uncovered. He talked about his feeling with others on the Executive Management Team and with three of the five NRC commissioners. But nobody really believed him—or wanted to.

It is hard to measure fuel damage without a sample of primary coolant water to check for fission products released by the exposed uranium dioxide fuel. And no sample was available on Wednesday afternoon. The auxiliary building and the plant's health physics lab were just too radioactive for a sample to be taken or measured.

Without a sample of the contaminated primary coolant, the amount of damage—the percentage of failed fuel—became an issue of intuition and debate. No one disputed that there was some failed fuel—that was the only possible source of the radiation in the plant. What was in question was the amount of fuel that had failed, the percentage of the fuel directly exposed to cooling water. The debate had definite economic undertones.

William Dornsife of Pennsylvania's Bureau of Radiological Protection estimated there was large fuel failure at 9 A.M. on Wednesday. Several of the Unit 2 operators mentioned feeling that there was a considerable amount of failed fuel on Wednesday. Yet, in his last statement of Wednesday, Jack Herbein confidently predicted that the amount of failed fuel was "considerably under one percent."

The NRC will allow a commercial nuclear reactor to operate with "up to one percent failed fuel." So Jack Herbein's prediction, made purely on intuition and maintained for two days, reflected the hope that the present core could be used and the $20-million replacement cost avoided. As the fuel-damage debate continued, at least 80 percent of the reactor fuel had failed.

Toward the end of the afternoon, it was decided that one of the main reactor coolant pumps should be turned on, to put the reactor in "a more normal cooling mode." For much of the afternoon, the system had been de-pressurized, with the Electromatic valve cracked open and high-pressure injection water throttled back in an attempt to force steam bubbles out of the system through the pressurizer. Running main coolant pumps with steam voids in the system still held the danger of "cavitation" and Met Ed was determined to save their investment in the main coolant pumps.

But there was a good chance, in the view of Gary Miller, that the afternoon's efforts had succeeded and one or more of the pumps might be able to start.

This feeling was supported by the NRC. At 6:45 P.M. Bethesda asked if Miller had considered restarting one of the pumps. This suggestion followed an afternoon of NRC inquiries that had been of questionable help. Following the accident, the NRC reviewed many activities of that first day and noted of its own actions:

"NRC's initial impact on the licensee [Met Ed] was to monopolize the attention of certain key licensee technical personnel in providing notification and communications. Licensee personnel involved in maintaining these communications were faced with difficult problems of nomenclature and spent much of their time training NRC personnel on plant systems. As time progressed, the nature of the communications shifted from training to information request and transmittal, and finally to information exchange."

Some of this "information exchange" included:

A 12:15-P.M. suggestion that the primary cooling system be "blown down" or de-pressurized—a maneuver followed by the utility and resulting in the reactor core's being uncovered for a second time.

A 2-P.M. suggestion that continued high-pressure injection would prevent the main core flood tanks from being activated, filling the reactor pressure vessel with cooling water. At the time, high-pressure injection flow was barely enough to keep the core from melting down, though the NRC suggested cutting it off completely. And activation of the core flood tanks with their 500,000 gallons of cold water might have destroyed what was left of the reactor core with thermal shock.

11

Wednesday Night

Night came to central Pennsylvania as a growing army of technicians worked to save the damaged plant. There would be no sleep at TMI. The plant was alive with lights and activity. Helicopters dropped from the darkening sky to land at the Island's heliport or passed overhead on radiation-monitoring runs. Higher still, radiation-monitoring aircraft passed overhead scanning the entire area at 8,000 feet.

The radiological-monitoring effort continued. Ordinarily only four monitors were required by Three Mile Island's license to operate. Now technicians distributed hundreds of dosimeters, hanging them in trees and on houses within ten miles of TMI.

NRC Region One health physicists had been taking off-site radiation readings during the afternoon. Their initial findings, despite frequent minor plant releases, indicated to them that there was no significant health threat. On-site levels "were not extremely significant, but higher than anything we had ever seen."

Maggie Reilly of the Bureau of Radiological Protection was delighted with the Department of Energy response. "They came into the office and then went out, doing their own thing," Reilly said. "One thing I appreciated with the federal response teams: when they came they didn't line up and say, okay, what do you want us to do; they just went out, you know, and grabbed a map and did stuff, which was far better than my wildest dreams in the past. The first line out of them was 'We report to you.' I was delighted."

But there was political infighting already going on in the federal agencies. Since midmorning, a special mobile laboratory had been driving up from a Department of Energy lab in New Orleans. By darkness, it was in Altoona, Pennsylvania, less than a hundred miles from Three Mile Island. This trip was totally unauthorized. The specialized lab had been sent from New Orleans because the director of that laboratory believed that the unit's unique capabilities and proximity would be important. The Environmental Protection Agency did not agree.

The EPA was put in charge of radiological monitoring to avoid friction between NRC and DOE, two essentially competing agencies in the area of nuclear power.

Instead of using the DOE lab from New Orleans that was practically at TMI, EPA called for a team to come in from its Las Vegas operation to take charge of high-level radiological monitoring. Their specially equipped Beechcraft King Air left for Harrisburg on Wednesday afternoon, followed by more than twenty technicians who traveled by commercial airline and by a helicopter that would take two days to make the 2,500-mile journey from Las Vegas to Harrisburg.

The New Orleans lab, sitting only a hundred miles from TMI, was told to turn around and go home. Las Vegas needed some obvious exposure to help justify a requested budget increase. Ironically, in this pointless competition between agencies, the EPA unit in Las Vegas was normally assigned to monitor DOE underground nuclear tests and the Department of Energy would continue to pay salaries and expenses of the EPA workers at Three Mile Island.

In Washington, Jessica Matthews was trying to gather as much information as she could about conditions at the site. Her source continued to be NRC Commissioner Gilinsky. At 7:30 P.M. she prepared another memorandum for Dr. Brzezinski, summarizing where things appeared to stand:

"The reactor itself is now under better control than it has been for most of the day . . . the two radioactivity meters inside the containment vessel are recording vastly different levels of radioactivity—10 and 6,000 rems per hour. [Normal reading is 5 to 10.] It is possible, though unlikely, that both are correct, since they are in quite different locations inside the vessel. The biggest problem we now face is to get inside the containment vessel in order to take actual samples. NRC has a team working on this tonight.

"The local utility has been in constant touch with the governor's office and with the Pennsylvania Civil Defense Council . . . The local congressmen and senators have been briefed by the NRC. In short, all seems under control on the political front.

"The cause of the accident is still unknown."

A newspaper reporter's professional life is spent mainly on the telephone, and the early evening of March 28 saw Gary Sanborn of the Allentown *Call* following true to form. Sanborn was trying to get technical information from Metropolitan Edison headquarters in Reading, without much luck.

"They were supposed to have some type of press center set up in Reading," Sanborn recalled, "but it sure didn't have enough phones. I tried for hours without getting through."

Sanborn was uniquely qualified among the reporters who covered the accident at Three Mile Island. For several years in the late 1960s and early 1970s, Sanborn had been a reactor technician charged with testing the cooling water of the eight small reactors that power the aircraft carrier USS *Enterprise.* He knew how reactors worked and he knew much of the language used by Met Ed—the technical language that confused so many reporters on the first day.

"I really didn't enjoy my work in the Navy," Sanborn remembered, "and I had tried to forget as much as possible about it. I went to college after being discharged and concentrated on journalism from that time on. Then this accident happened and I really had to stretch my mind to remember."

For every Sanborn who leaves his Navy nuclear training behind, however, there are several other veterans who go on to work in one of the nation's commercial nuclear power plants. Gary Sanborn had friends who had stayed in the nuclear industry and he called as many as he could on Wednesday.

"I needed information about how the plant worked and they gave it to me. It was odd, too, that here they were, working in the nuclear industry and they were relying on me to give them information about what was happening at Three Mile Island. One friend who works for a utility in New York told me to find out whether the emergency core-cooling system had been activated, whether there was evidence of fuel failure and other important questions that we were the first to ask. I fed them on to Ben Livingood in Harrisburg, then tried to get a source inside Met Ed."

In the meantime Ben Livingood was also trying to get Met Ed headquarters on the phone. "You just have to be stubborn," Livingood said later. "I sat on the telephone with the flacks in Reading saying, 'Give me someone who can answer my technical questions. I don't want to talk to you. Give me someone who knows what's going on.' Gary was doing the same thing from Allentown, but I was the first to get through. Dave Klusick finally connected me with a guy named Mike Buring. What we wanted was a sequence of events —just what happened in those first hours of the accident after 4 A.M. He was helpful and gave us everything he had in time to make our Thursday deadline. We were the first paper to give a chronology of the accident—beating out the rest of the country by a couple of days. And we'd found a source inside Met Ed."

Sanborn and Livingood had a sequence of events that included the fact that the Electromatic relief valve on the pressurizer had failed —a point that had not previously been released and would not even be mentioned when Jack Herbein briefed the press the next morning.

If the Met Ed public information office in Reading seemed confused to Livingood and Sanborn, it was. David Klusick had been answering press and public telephone questions along with two other communication services staff people, Judy Botvin and Don Curry. But three people just weren't enough.

". . . I think it was Wednesday, I'm pretty sure it was Wednesday afternoon," Klusick recalled later, "some of the employees from surrounding departments—notably rates and consumer services— knew that we were just swamped and whether they asked or they just filtered in themselves, I don't know. I still don't know, but they asked if they could give us a hand, and we said 'You bet.' We gave them whatever statement we had at the time. We couldn't give them any real briefing on who was calling."

The word on every caller's tongue that night was "meltdown." Klusick remembered: "It was reporters, it was members of the public, it was some co-workers who may have heard a radio broadcast or a television broadcast . . . And these guys on the phone were just about having a heart attack, some of them. I mean, you know, they would fire off their question, you could barely understand it, and you try to stop and say, 'Let me explain what you're asking and make sure we're talking about the same thing.' Man, they didn't want to hear whether you were talking on the same frequency, they wanted —I don't know what they wanted to hear. I was trying to really make sense of their question but, my God, if you couldn't give them a two-sentence answer, it was goodbye and I'll write what I damn well please."

Dave Klusick said, again and again that Wednesday night and throughout the next week, "This is *not* a China Syndrome situation!"

While Met Ed was trying to mount a response to the accident, the last thing the utility wanted was for reporters to show up at the company headquarters in Reading, and they made this clear to the press on Wednesday. But while other reporters were jamming Met Ed phone lines in Reading and Hershey, Dan Machalaba of *The Wall Street Journal* was taking advantage of the confusion to walk into the building for a few interviews. The enterprising reporter followed a Met Ed staffer to the third-floor offices of Met Ed president Walter Creitz. He interviewed Creitz and George Troffer, head of quality assurance for the utility, and, as a Met Ed accident chronology put it, "stayed well into the evening observing the activity."

While Machalaba was stalking the hallways of Met Ed, the Philadelphia *Inquirer* was taking a second look at its own coverage of the accident. The *Inquirer* had a two-man bureau in Harrisburg and had relied mainly on that bureau and the wire-service reports for the first day's coverage.

But now it looked as if Three Mile Island was going to be a big story, and so the *Inquirer* began to devote the kind of resources to the accident that had gained the paper five consecutive Pulitzer prizes. Wednesday night, the Philadelphia *Inquirer* sent more reporters to Harrisburg than any other news organization—thirty altogether. A reporter from another paper observed, "the *Inquirer*'s people were falling all over each other."

Three groups were sent to Harrisburg: a core unit of reporters and photographers to cover the central story of what happened and two investigative units intended to look into the history of the plant, interview as many plant workers as possible and put together a definitive account of the accident. The investigative groups were put nominally under the control of Rod Nordland.

"Nordland is the sort of reporter who wouldn't have made it on a lot of other papers," *Inquirer* metro editor John Carroll said. "He didn't care about things like appearance or making a good impression, and if he didn't like a story assignment, he often didn't do it. Most papers wouldn't have put up with that kind of behavior. At Three Mile Island, I don't think he even slept."

While the *Inquirer* forces were in the field in Harrisburg, Middletown and Goldsboro, Susan Stranahan stayed in Philadelphia to write the main accident story for each edition. She found Nordland to be a colorful character too:

"He used to look like a member of a motorcycle gang, had hair nearly down to his waist that he wore in a ponytail and always seemed to be dressed in black leather—jacket, pants and all those metal studs. Imagine a character like that asking questions of Frank Rizzo at a press conference! He did some stories on motorcycle gangs too, and one time they beat him up. He's really crazy, insanely devoted to a story if he likes it. He'll do anything, stay up for days and never run out of energy. But that's if he likes the story."

Shortly before the accident at Three Mile Island, Rod Nordland had been assigned to cover the Pope.

In New York, Jonathon Ward of CBS was trying to get the measure of the accident at Three Mile Island. He had committed two crews to cover the scene of the accident and had Bob Schakne working the agencies in Washington. Was the story going to get bigger, or was it already past its peak?

"I tried to get information from all sides," Ward said. "First, I called the Union of Concerned Scientists in Massachusetts. They didn't have very much on the plant or the accident. Then I tried the Atomic Industrial Forum in Washington. They are so thrilled when I call them for information *before* doing a story. This time, though, there was a lot of confusion and the last thing the guy I talked to said was 'This whole fucking thing was produced by Jane Fonda!'"

Ward decided that the accident was an even bigger story than he had thought at first. That night, he packed a suitcase to take in to work the next morning . . . just in case.

Wednesday night a new film started playing at the Camp Hill Theater across the Susquehanna from Harrisburg. Called *The China Syndrome,* it starred Jane Fonda, Michael Douglas and Jack Lemmon. It was a film about nuclear safety and about television.

Evelyn Hipple worked at the candy counter in the movie theater. She could see the films for free, but only watched the first half of this one. "Oh, it's too upsetting," she said.

The film was especially upsetting to the nuclear industry. One executive of Southern California Edison said the film "hasn't any scientific credibility and is, in fact, ridiculous. *Syndrome* rendered an unconscionable public disservice by using phony theatrics to frighten Americans away from a desperately needed energy source."

Film critic Charles Champlin found the film to have "a potently proposed point of view and it is not ignorable." No one in central Pennsylvania that night was ignoring it.

Paul Hunter came out of the theater and said, "You take the combination of the movie and the event, and it brings it right down to your doorstep. Like they say, the plant could blow straight to China."

"It's too drastic even to think of promoting business on the basis of this," said theater manager Spike Todoroff. "I must admit though, it's a godsend that this movie is playing at the time this happened. It's going to help business. It is helping business."

"I find it disturbingly ironic," said Michael Douglas, the film's producer. "All we tried to do was promote the movie as a thriller.

We didn't ask for the controversy. Some critics said we were trying to alert the public for no reason at all. But then science editors began seeing the film and realized that we had done our homework."

To many living around Harrisburg, *The China Syndrome* was the only way to understand just what was going on at Three Mile Island —a way to justify their fear. Days later machinist Joe Buela saw the film, saying, "I just want to find out as much about it as I can. Your town makes the covers of *Time* and *Newsweek,* you want to find out why."

The two accidents depicted in *The China Syndrome* were very different from what happened at Three Mile Island: The control room of the movie's fictitious Ventana Nuclear Power Plant was much more modern, better designed and better thought out than the control room of TMI's Unit 2; the accident at Three Mile Island was far more serious than either of the two transients portrayed in the film; the film expounded a theory that had been popular among nuclear experts until it was proved wrong that Wednesday morning at great expense to Metropolitan Edison and the electricity users of central Pennsylvania:

GREG GILBERT [reactor expert in the film]: . . . I'm not sure, but they may have come very close to exposing the core . . . If the core is exposed, for whatever reason, the fuel heats beyond core heat tolerance in a matter of minutes. Nothing can stop it. It melts then, right through the bottom of the core, the container, through the concrete basin and, in fact, it melts right through the bottom of the plant.

At Three Mile Island, that Wednesday, the reactor core of Unit 2 was uncovered for hours, not minutes. And something *did* stop the core from "melting through the plant"—but left half a billion dollars in damage.

106

12

A Cloud of Radiation

Paul Critchlow had come to work for Richard Thornburgh from the Philadelphia *Inquirer,* where he had been a political writer. He quit his job to join the campaign of a candidate for governor thirty points behind in the polls. It was a big risk on Critchlow's part, but one that paid off. And Thornburgh realized what a risk Critchlow took. As a result, the governor's press secretary gained more influence than is usually expected for one in that position. As the close personal adviser to a governor who trusted few of the people under him, Critchlow was probably the third most powerful man in the government of Pennsylvania. He had far more clout than did Lieutenant Governor Scranton.

With only unreliable information coming from Met Ed, Thornburgh put Critchlow on the job of finding out what was really going on. In this capacity, he and his staff became more information-gatherers than disseminators. In fact, the function of keeping the press informed virtually stopped as Critchlow and his staff devoted all their time to gathering information about the accident for Thornburgh. This change alienated the press, who found the press office to be almost useless to them, and it ultimately proved to be a disservice to many citizens of Pennsylvania, who later called with questions or rumors for confirmation.

The first thing Critchlow needed was accurate technical information from the site. For this, he turned to the nearest of five NRC regional offices. Shortly after 7 P.M. three inspectors from NRC

107

Region One—covering the Northeast—were called to Harrisburg to brief Lieutenant Governor Scranton and others in the state government.

Charles Gallina, Jim Higgins and Don Neely had represented the NRC at Three Mile Island for most of the day. In addition to observing in the control room and transferring information to the Incident Response Center in Bethesda, the men had been taking off-site radiation readings during the afternoon. Their initial findings, despite frequent minor plant releases, indicated to them that there was no significant health threat. On-site levels "were not extremely significant, but higher than anything we had ever seen." Their attention was focused on stopping the intermittent releases of radiation.

"During the whole day on Wednesday," Gallina said, "we had absolutely no idea of any of the incidents that led to the accident itself. You can appreciate walking into a control room. You have a reactor that is in an unstable condition . . . not in any position or any status that you have normally seen it before. It is not operating. It is not in cold shutdown, pressures are changing, temperatures are changing. They had voids in the system they didn't know how to get rid of, other problems that didn't jibe with what you normally would expect.

"Our primary concern was 'Let's get the thing stable.' Anything related to what in the hell caused this thing or who did what to who had to take a back seat. Once the plane crashes, then the FAA goes out and tries to find out why, but here we have a situation where the plane hadn't crashed. A DC-10 had lost an engine and the pilot is trying to keep it flying with two remaining ones. Okay, we want to land the plane. Once it is down, we can start looking into why the engine fell off and why we are having trouble controlling it, but at this point we don't really care about that."

Despite their busy afternoon and the fact that there was much left to do, the NRC inspectors decided to leave Three Mile Island more than an hour before their scheduled briefing of the lieutenant governor at the state capitol only nine miles away. Radioactive xenon gas had been wafting through the control room all day, and they probably had some of it on their clothing. By leaving for Harrisburg early, the three NRC men were allowing time for any gas on their clothing to decay away—lose its radioactivity—a change that usually takes less than an hour.

Entering the observation center, sure enough, the frisker went off, showing all three to be covered with the radioactive gas. Time passed and the Pennsylvania state policeman waiting to drive them to the capitol said they should go. Gallina and Higgins departed for Harrisburg, leaving behind radiation physicist Don Neely, who was still contaminated, a victim of his own taste in clothing.

"It tends to stick to your pants," Gallina said, "especially to polyester double-knits."

When Higgins and Gallina arrived at Scranton's office on the second floor of the state capitol's executive wing, they were met by Scranton and his aide, Mark Knouse, Colonel Oran Henderson of PEMA, Tom Gerusky of the state's Bureau of Radiological Protection, Robert Friess of the Department of Energy, and Governor Thornburgh's assistant, Jay Waldman.

When the meeting began, Higgins and Gallina still wore their hard hats. They gave an honest assessment of conditions at the power plant, as it seemed to them. This was important to Scranton and Thornburgh, who had lost all faith in Met Ed by this time. The men meeting in Scranton's office knew that the press would soon be telling how George Troffer of Met Ed said radiation levels inside the plant were ten times normal, while Edson Case from the NRC Incident Response Center in Bethesda said they were a thousand times normal. Someone had to be wrong, and Jay Waldman was determined to find out who that was.

There were great similarities between Jay Waldman and Richard Thornburgh. Both men were lawyers and both had been successful prosecutors. Thus they gained information mainly through questioning and cross-examination and tended to doubt what was told them. This produced a highly accurate but slow method of administration for the Commonwealth of Pennsylvania. Decisions were reached only after questioning and re-questioning. Jay Waldman's function, during the accident at Three Mile Island, "was in the role of trying to ask relevant questions, to flush out relevant information, identifying options and consequences, present them cogently to him [Thornburgh], advise him when he wants advice, reach an equitable decision, just like I do every day."

Waldman was true to form at the briefing in Scranton's office that night:

"I remember asking them pointed questions to try to cut through the technical jargon. I remember asking them to please explain in simple English terms what the hell happened there, what could happen, what were the probabilities that each of these options could develop, and what kind of time did we have? I recall that it was quite difficult to get them to answer questions like that for a while."

Gallina was encouraging. He told the state officials that he suspected there may have been some exposure of the core during the accident, that the worst possible event was a meltdown, but the probability of a meltdown was extraordinarily low and that 20 to 30 hours would be available in which to react. Most especially he stressed that there was no substantial risk or danger off-site and no permanent damage to the plant.

Chick Gallina was mistaken.

"We knew that the plant, in going through the transient, had experienced . . . a bigger insult to its integrity than it would normally have experienced," Gallina later said, ". . . but in our minds at that point, we were thinking of the failed fuel as hairline cracks, pin holes, maybe some tears, stuff like that which had released activity to the coolant. We needed a primary coolant sample to really know the damage, and we didn't get one until the next day."

The briefing gave Scranton more confidence in the way the accident was being handled. Scranton asked Gallina, "Would you mind talking to the press?" All the participants in the briefing assembled with the press in the cramped capitol media center. It was Scranton's third press conference of the day. The media center, designed to hold fifty reporters at most, held at least a hundred. Bodies crushed together and helped the television lights to make the room unbearably hot.

Gallina and Higgins presented the same optimistic outlook to the press that they had to the lieutenant governor.

110

"There was some equipment failure," Higgins said in reply to a question about permanent damage to the plant, "and there was some damage of a minor sort because of the transient that occurred today. [But] as far as permanent damage, nothing that can't be repaired with normal maintenance."

"At this point in time, we have been investigating it most of the day," Chick Gallina said. "Of course we haven't gotten into a detailed investigation because we are getting the plant into a stable condition, but there has been absolutely no indication of human error at this point." The press was told that there was no significant core damage and that the "reactor should be in cold shutdown within a day also."

Ben Livingood was having trouble hearing the answers Scranton and the experts from the NRC were giving, much less getting in his own questions. But he did manage to ask a few, the most important being: "Was the emergency core-cooling system activated?"

Gallina hesitated before giving a long roundabout answer to the question: "A lot of the systems on-site are dual usage. That is, they are used for emergency core-cooling and for normal plant operation. Some of the pumps are used for normal makeup and injection concurrently, they can be used if you have an accident, for emergency core-cooling. The situation that classically presents the need for emergency core-cooling is a pipe rupture where the system is depressurized and you have to inject water with these emergency pumps into the plant. That situation did not occur. The pumps, the same pumps were being used in the normal mode, to inject water and to bleed water off from the plant system and that has been done throughout the day."

Gallina was inaccurate when he said that the plant had not been de-pressurized, but right that the emergency core-cooling system had not been used to maintain pressure in the primary cooling system. It should have been used that way, but Craig Faust had turned the system off.

Still, Gallina had admitted that the emergency core-cooling system had been activated. Gary Sanborn had suggested the question to Livingood, who did not really understand what he was asking.

111

When Livingood reported Gallina's answer to Sanborn that night, "he told me that it meant we had come pretty close to something like a core meltdown."

That evening, Middletown's Mayor Bob Reid was cruising the streets of his town with Earl Anderson, a Middletown policeman. The city council held most of the power in Middletown and Mayor Reid's major responsibility was public safety. He took the responsibility very seriously. So, every Monday, Tuesday and Wednesday nights, from seven to nine, Reid rode with his friend Anderson. They talked about sports or what was in the news and Middletown could see that its mayor was on the job.

After nearby Olmsted Air Force Base closed in 1964, the start of construction at Three Mile Island had brought needed employment to the men of Middletown. And the plant was much cleaner than Met Ed's old coal-fired Crawford plant that had stood for decades at the edge of the city, spewing smoke and soot. But that afternoon, Reid had watched the needle on his Civil Defense Geiger counter move in harmony with bursts of radiation from Three Mile Island. Radiation! This was a new and unexpected threat—one that Mayor Reid wasn't sure how to deal with.

Only radiation really wasn't the novel threat to his city that Bob Reid thought. Middletown had been exposed to radiation before without ever knowing it. During the late 1950s—at about the same time Dr. Edward Teller said to a questioning Groucho Marx on national television, "How do we know? Fallout might be good for us!"—a nuclear weapons accident at Olmsted Air Force Base laid a big cloud of radiation right through Middletown. It was the cold war and what harm could a little cloud of radiation do in a small city in Pennsylvania? So the public was never told of the incident.

Months after the accident at Three Mile Island, the U.S. Department of Health, Education and Welfare and the Pennsylvania Department of Health would commit millions of dollars to complex longitudinal studies of public health around Three Mile Island. Their task would be to statistically isolate cancer cases caused by the accident in Unit 2. The politically motivated studies, considered by

112

many health professionals to be a waste of money anyway, were made totally worthless by the unknown effects of the cloud of radiation that had floated through Middletown on its way from Olmsted Air Force Base in 1958.

Watching the flickering needle on his Korean War–vintage Geiger counter, Mayor Reid was becoming angry. Damn it, they should have been warned, they should have been told the truth. It had taken four hours for word of the accident to reach the three miles to Mayor Reid—and even then he had heard from Civil Defense, not from the nuclear plant. Met Ed was planning a press conference in Hershey for the next morning. Driving back to the town hall with Earl Anderson, Mayor Reid decided that he would be at that press conference to get a good look at his enemy.

The Governor's Mansion stood behind an iron fence, facing the Susquehanna River on the far side of Harrisburg from Three Mile Island. The sprawling tan-colored brick building was bathed in light, both for publicity and security. Two blocks away, a gas station stood, burned out by an arson.

It was mild for a Wednesday night at the end of March, and many people were out enjoying the evening along the river. Joggers and cyclists shared a lighted path along its shore.

Thirty people were gathered in front of the Governor's Mansion. Some looked across the fifty yards of grass to lighted windows in the main building. Others carried signs and watched where they were marching. A pregnant woman carried a cardboard sign wrapped across her swollen belly: SHUT DOWN TMI!

Governor Richard Thornburgh left his desk at 10:40 P.M. to talk to the people gathered in front of the house he had lived in for just over two months. It was a rare informal appearance for a governor whom some in his own office refer to as "a cold fish."

Thornburgh was reserved, introverted and intense. His appearance was generally conservative, to match many of his views. Black shoes, gray two-piece suits and horn-rimmed glasses dominated his wardrobe. After coming from behind to win the governor's race, Thornburgh showed a rare example of his humor when he said that

113

his election had "made the world safe for horn-rimmed glasses."

This night, the trim, dark-haired governor walked out in shirt-sleeves to spend a few minutes with those who had come to his house. As always, a state policeman went along.

The governor's line was one of reassurance: "We have the best experts in the country here working on the problem . . . I have been told that the situation is improving . . . We don't expect it to get worse, though we would be ready in any eventuality . . . No, I'm not sure I have the authority to shut the plant, even if I wanted to . . . We'll know more tomorrow."

Two cars turned in the mansion's side gate. One was a state police car. The other belonged to Lieutenant Governor Scranton, who was accompanied by Higgins and Gallina. Thornburgh took them inside to learn the latest word from TMI. The pregnant protester removed her sign and put on a coat.

Chick Gallina told Thornburgh the same story he'd told Scranton and the press earlier in the evening: The situation is getting better, no permanent damage to the plant. Thornburgh asked pointed questions.

By the end of the briefing, the governor was satisfied that a consensus existed as to conditions at the plant. "There was no sense of urgency about the steps that had to be taken," he said.

But while there seemed to be no urgency in Harrisburg, the same was not true in Bethesda, where retired Air Force Major General Lee Gossick finally had his Incident Response Center running smoothly with a full staff and adequate communications with Three Mile Island. The accident was lasting longer than anyone had expected; the reactor was cooling far more slowly than anyone in Bethesda thought it should.

At the end of Wednesday evening, there were seven health physicists and four reactor inspectors representing the NRC at Three Mile Island. The contradictions in statements and apparent information had made it obvious to Gossick that more NRC people were needed in Pennsylvania. And it bothered him that the people on-site under Gallina were all from Region One—there was no one from Washington.

By early evening Harold Denton had arrived in Bethesda. The head of the Operating Reactor Division of the NRC had been busy on other matters for most of the day. He had planned to be in Phoenix that night and in California on Thursday for more hearings on the controversial Diablo Canyon nuclear plant, but these plans were changed quickly.

Up to this point, Denton had not thought he would be needed, "knowing these I&E [Inspection and Enforcement] gumshoes would come through eventually and establish exactly who struck out on this." But Denton soon came to share Gossick's concern about Met Ed's problems in coping with the accident and he decided to send a group of experts from Bethesda to help handle the growing problem.

Dr. Richard Vollmer was selected to lead the team from Bethesda. Vollmer was head of the NRC's research branch and knew as much as anyone about how reactors worked . . . in general. But, like everyone else in the NRC, Vollmer was not licensed to operate a nuclear power plant. He went home to get ready for the trip to Harrisburg in the morning. Denton spent the night on a cot in Bethesda.

Late that same night, Richard Thornburgh was looking for a book. Several years before, he'd read *We Almost Lost Detroit,* a book about the near meltdown at the Fermi Nuclear Power Plant in Michigan. The governor remembered "the ghastly scenario that was laid out about the core damage." He also remembered that he had not asked Gallina and Higgins about core damage at TMI. Not finding the book, Thornburgh went to bed, intending to ask the experts about it in the morning.

13
Thursday

At 5 A.M. Thursday morning, Kevin Molloy crawled out of the Army cot that had been his bed for what was to be the first of eleven consecutive nights. Only a couple of hours before, the Dauphin County Civil Defense director had crawled onto the cot in his office under the county building in Harrisburg. Wednesday had been a twenty-hour workday for him. Thursday would be just as long. And Friday would be longer.

Padding about in stocking feet, Molloy went into the communication center that controls most of the county police and fire companies. No one was asleep there. Communication had been Molloy's major concern on Wednesday. He had local Civil Defense directors, mayors, county commissioners and citizens to keep informed in the area surrounding Three Mile Island. Getting current updates on the situation at the plant and relaying them by phone, teletype and radio to those who should know had taken much of his time.

None of the communities within five miles of TMI had an evacuation plan in case of an accident at the nuclear facility. Met Ed had said five years before that no plan was really needed. Molloy did not agree. However, recognizing a need and getting local volunteers and politicians to respond were two different things.

"We had a meeting in November," Molloy said, "and I invited the local police, fire and Civil Defense people from the five-mile zone up here. And we went over basically what could happen if an incident

116

occurred at TMI and what their responsibilities were. And I suggested—everybody knew what to do, but I said, you know, we ought to put it in writing.

"The old state and federal laws simply require that everybody have a Civil Defense director. There is no training requirement—no nothing . . . There are incidents where in the five years I have spent here, prior to TMI, where I have had no local directors come to our training sessions. You know, in five years they never attended one meeting."

Mayor Bob Reid decided long before the accident at Three Mile Island that his town needed an emergency plan.

"I tried to get everyone together to write up a plan," Reid said. "We had a meeting and a lot of people showed up. There was a lot of interest at first. Then we had another meeting and fewer people came. By the third meeting, it was just Butch Ryan, George Miller and me. We were still trying to put together a plan when things started to happen out at Three Mile Island."

The events of Wednesday had done a lot to make up for years of local indifference to emergency planning, and Molloy had been kept busy advising Civil Defense directors and police chiefs on what to include in the plans that were finally being written. It took a nuclear accident and a lot of public fear to bring emergency preparedness to Dauphin County.

But what had seemed so bad the day before looked better in Harrisburg that Thursday morning. At 5:45 A.M., Molloy called PEMA to get the latest story on conditions at the plant. The news was very good, with the reactor under control and far more stable than it had been the night before. The accident might be over and with nobody hurt.

Molloy and his dispatchers began to spread the good news to all the local officials who relied on them as the only official source of information about what was happening at TMI. Most of it came directly from Colonel Henderson, the head of PEMA, who had been involved in all the briefings the day before and participated in all three of Scranton's press conferences.

117

* * *

The press had spent much of Wednesday getting the measure of Three Mile Island—deciding whether or not it was an important story and then learning enough about nuclear power to report what was happening. By Thursday morning, all the cramming and effort began to pay off in extensive and comprehensive coverage of the accident in the national media. Most of what was said was encouraging.

Met Ed President Walter Creitz appeared that Thursday morning on NBC's *Today* show. Creitz avoided the broad question of operator error and tried to point out the industry's safety record:

> CREITZ: Keep in mind we have seventy-two nuclear reactors in operation in this country, some of them since 1954. And we still haven't injured a single member of the public. And I think that's significant.
>
> There's no technology that man has ever been involved with since the first time he started to use fire that can claim a record like that.
>
> TOM BROKAW: You mean there was no human error involved at all?
>
> CREITZ: There was no human error involved in any of that procedure.

While Creitz was appearing on NBC, over on ABC, Harold Denton of the NRC was being interviewed on *Good Morning America.* Denton tried to look at the accident as though it was ending, but admitted it was "one of the worst in U.S. history."

In contrast to Creitz and Denton, there was much more critical talk coming from Robert Pollard of the Union of Concerned Scientists on the CBS Morning News. Pollard, a former NRC official who resigned over nuclear safety issues, made a number of candid statements that morning:

> POLLARD: There were a combination of mechanical failures and apparently also some human errors that brought us at least partway down the road to a major meltdown. It's impossible to say at this time exactly how close we've come, but damaging the fuel is the first step toward the process of having a major catastrophe . . . The event that started all of this is a routine event. It should have had no effect whatsoever, inside or outside the plant . . .

118

What I would say is to take this as a warning. It does not establish that we know that the safety systems are adequate . . . Unfortunately, it confirms to me that my reasons for resigning were correct. That is, the Nuclear Regulatory Commission [is] licensing plants without an adequate view. In this accident at the Three Mile Island plant, we've obviously had multiple failures of some of the safety systems. This plant has been plagued with failures of the very same systems since it was first licensed in February of last year, and apparently nobody's heeded those warning signs . . . this particular plant has affecting it at least 22 different serious safety problems which are among 133 different problems that the NRC reported to Congress in January of last year . . .

Watching this interview at his Connecticut home, Jonathon Ward of CBS decided it was time he left for Three Mile Island.

Those antinuclear spokesmen who were making statements on Thursday were well covered by the press: their remarks were simple, direct and strong. But many reporters had their own reservations about the antinukes.

Jim Panyard of the Philadelphia *Bulletin* said, "The problem with antinuclear groups is that they have been crying in the wilderness for a decade, and the only way that they have been able to get attention is by telling about the worst possible nightmare. I would want the answer to a specific technical question such as 'Could this breach the containment walls?' They would answer, 'Yes, that could split the containment and then thousands of people will be killed.' They couldn't forget the nightmare and answer a simple question. I was thinking of the next twenty-four hours and they were thinking of the future of mankind."

Congress may not have done anything about the 133 safety problems cited by Pollard, but there was great interest on the Hill that Thursday morning in what had happened at Three Mile Island.

Constituents had been calling Pennsylvania Senator Richard Schweiker's office since midafternoon on Wednesday. Many callers just wanted information about what was happening at Three Mile Island. But other callers, according to Schweiker's legislative assistant, Dwight Geduldig, "wanted the senator to go right in and shut down the nuclear plant."

Schweiker was one of a number of senators and congressmen slated to tour the plant that afternoon. Led by Colorado Senator Gary Hart, the helicopter trip was billed as a "congressional fact-finding expedition"—an expedition made harder by the fact that Bernie Weiss of the NRC still had not yet found rental cars in Harrisburg for the group.

While Gary Hart was mustering his forces for the trip to Pennsylvania, Congressman Morris Udall had hastily scheduled a hearing for his energy subcommittee to examine causes and management of the accident. And the commissioners of the NRC were doing some fact-finding of their own in an attempt to decide just what had happened in Pennsylvania the day before and what should be the function of their organization in responding to the emergency.

As a successor to the old Atomic Energy Commission, the NRC suffered a major identity crisis. The AEC had been both nuclear power's regulator and its booster, but a 1974 reorganization supposedly left the boosting to the Department of Energy (then the Energy Research and Development Agency) and the regulating to the NRC. Still, most of the NRC personnel had come over from AEC.

When asked how long he had worked for the four-year-old NRC, Darryl Eisenhut replied "Eleven years." Eisenhut began work for the Atomic Energy Commission in 1968.

And the NRC was having a difficult time getting its public information effort into action. Karl Abraham was on the scene in Harrisburg and not being especially helpful to the press. In Washington, NRC head of public information Joe Fouchard was bearing the brunt of reporter inquiries.

Peter Hackes of NBC said of Fouchard: "You could ask the most piercing, searching questions and get the most piercing, searching I don't knows. That made me shaky. I didn't know who to ask."

Reporters who covered the regulatory commissions as a normal beat were not surprised at this sort of behavior from the NRC. It was a small agency, not used to public attention and operating erratically while in the limelight.

Robert Ruby of the Baltimore *Sun* said, "I've been to NRC meetings, and there are so few people there you could shoot a cannon through the room."

Even stronger words came from another regulatory-beat reporter who said of the NRC after TMI: "It's a dinky little backwater asshole agency that will drop back into obscurity when this is all over."

Lee Gossick, who would be the first NRC man to lose his job because of Three Mile Island, began the presentation to the commissioners that morning in the commission meeting room on H Street in Washington. Gossick told them about the NRC contingent at work in Pennsylvania and then turned to Darryl Eisenhut for an assessment of the technical situation. Like most other assessments of the damage that morning, Eisenhut's was sanguine. "We still didn't have a primary coolant sample," he said later in explanation.

As the meeting continued, a picture of the accident emerged that was without detail or obvious direction, though the commissioners did not notice this at the time. Relieved that the worst seemed over, they let much of what Gossick and Eisenhut said wash over them, unchallenged.

Eisenhut admitted a lack of basic information about the cause of the accident. Questioned by Chairman Joseph Hendrie on the initial loss of feedwater, he replied, "We know no details on the actual . . . the extent, the sequence, the timing, the degree to which we lost the main feedwater."

Hendrie could have gained more understanding of the accident from that morning's Allentown *Call,* where Ben Livingood and Gary Sanborn presented the first accident chronology. Thursday was the first day Hendrie had been actively involved in details of the accident, Wednesday having been involved with attending to his twenty-one-year-old daughter who was having oral surgery.

Eisenhut expressed some reservations about Met Ed and the handling of the accident—"There has been enough anomalous behavior that we don't want to speculate too much"—but held back from recommending that the NRC take over management of the accident.

There was some concern among the commissioners that this was the logical move—to take over the plant and bring it to a cold shutdown as soon as possible. Gossick suggested that Vollmer and the group of NRC experts leaving for Harrisburg that morning would ". . . literally assume responsibility for seeing that the state of the plant is kept in a safe condition." But Gossick had to admit that his people could not actually take over running the plant. There were no licensed operators available. The NRC was less able to run Three Mile Island than the utility it was supposed to oversee.

The scenario Eisenhut put forward for cooling down the plant was already a familiar one: de-pressurize the reactor until the core flood tanks were activated, filling the core with half a million gallons of cold, borated water. But Eisenhut didn't mention that this method had already been tried twice—both times without success.

After the meeting Hendrie went to Capitol Hill to report to Representative Morris Udall's energy subcommittee. The fifty-four-year-old physicist said the trouble was over. "The way he was talking," a congressional staffer said later, "the whole thing amounted to little more than a midget's breaking wind."

The Hershey Motor Lodge was a posh resort. It was owned, like almost everything else in town, by Hershey Foods, the world's largest chocolate maker, and sat with its tennis courts, pools and golf course right next to the Milton S. Hershey Medical Center, owned by Pennsylvania State University but paid for with 80 million Hershey dollars. It was nine miles from Three Mile Island.

Late that morning the Hershey Motor Lodge's gold-carpeted Aztec Room was the site of Met Ed's first official press conference since the accident. Jack Herbein and Walter Creitz, representatives of a utility that was strapped for money, cutting staff and plant maintenance and now facing a multimillion-dollar repair bill at TMI, chose to meet the press at one of the most expensive resorts in Pennsylvania.

By 11 A.M., more than 120 newsmen filled the room. A dozen

122

television crews, including all three networks, trained their cameras on a podium set with 40 microphones. In the middle of all this sat Middletown's Mayor Reid and his chief of police, George Miller. "Whoa, this is big stuff," Reid whispered to Miller. "And here we are, two country boys."

The press was happy to have a news event to cover that morning, because little seemed to be happening out at the Island. Met Ed and the NRC had become more organized at the site and it wasn't as easy to get a story as it had been the day before. Chicago *Tribune* science writer Casey Bukro said, "After a while they had ropes and guards around the observation center. You had to ask for the person you wanted to see and they'd send a runner."

Herbein and Creitz stood in front of an American flag in the sweltering room. Both men looked tired. Herbein wore the same clothes in which he'd met the press at the TMI observation center twenty hours before. Contradictory statements by Met Ed and the NRC had created an adversary relationship between the utility and the press. Creitz and Herbein were in for a difficult hour.

Herbein made the opening statement, confident he was on top of the technical situation: "Well, this morning we've got our reactor cooling pump running. We're removing heat from the core. The plant's in a stable condition and, we believe, operating normally. Later on today, we would hope to perhaps shift over to our decay-heat system. And at that time the plant would be in the cold-shutdown condition."

From that point, the press conference went downhill for Herbein. Peter Stoler of *Time* magazine pointed out that Herbein gave the same core temperature as the day before, yet said the core was cooling. How could this be? No answer.

Was there core damage? "What we've seen here is a fuel failure, something that's possible with an operating reactor today, and no-body's ever said this couldn't happen. The safety systems were designed to take care of this kind of accident, as our safety systems did."

"What's not possible," Creitz said, "is the hypothetical accident you hear about in which the entire core of the reactor melts and spews molten radioactivity into the area for miles around and kills tens of thousands of people. That's what we've been telling you is not possible."

Pressed by the assembled reporters, Herbein and Creitz estimated that less than one percent of the fuel rods had failed—still within NRC standards for reactor operation. They did not deal with the possibility of operator error and didn't mention failure of the pressurizer's Electromatic relief valve.

"I can tell you that we didn't injure anybody through this accident," Herbein said. "We didn't overexpose anybody, and we certainly didn't kill a single soul, and as I've indicated, levels of radiation which re—which were released off-site were absolutely minuscule."

While the reporters were shouting their questions, Robert Reid sat in his pin-stripe suit, sweating in the overcrowded room and getting increasingly angry. His constituents were the ones put out by all this, they were the ones who were not told of the accident until it might have been too late. Reid, who was supposed to be giving a test to his government class at Middletown High that day, jumped to his feet and grabbed the microphone in the middle of the room—the microphone ignored by the shouting reporters.

"I'm Mayor Bob Reid of Middletown," he boomed. Three hundred eyes and nearly as many lenses turned to face Reid. "I thought I was going to get mugged," he said later.

Reid wanted to know why it took more than three hours for Met Ed to tell the world about the accident. "We were the ones most affected and yet we were the last to know. We had no way of knowing what the extent of the danger was. And right now we still have no way of knowing. I'm upset and angry."

Jack Herbein launched into a typically complex technical answer to Reid's emotional outburst. It didn't work. Creitz watched his boy-wonder vice president floundering, yet Herbein didn't seem to realize he was floundering. Creitz saw how angry Reid was and how taken the press was with it all. Stepping to the microphone, Creitz stopped Herbein in midsentence and made a statement to Reid that was unique in Met Ed's relations with the press and handling of the

124

accident. Looking out at the black mayor, standing defiantly with his feet apart in the crowd of reporters, Creitz could only say, "We're sorry."

But Mayor Robert Reid didn't hear. Still fuming, he was on his feet marching out of the room. And most of the press was going with him.

14

The Dumping

At 11 A.M. a dozen aircraft sat waiting to take off from Teterboro Airport, across from Manhattan. The gleaming business jets, heavy twins and turboprops, holding their cargo of buttoned-down, three-pieced, high-priced corporate talent, cost up to $5 a minute just waiting their turn to take off.

Elbowing his way into the monotonous radio traffic of request, clearance, request, clearance, the last pilot in line broke the ritual of departure.

"Uh, Teterboro Ground Control, we'd like to, uh, expedite departure if we may," the pilot radioed. "This is an emergency flight carrying a WCBS news crew to Harrisburg, Pennsylvania."

"Roger," the ground controller replied. "That's up to the traffic ahead of you. Will you let the Channel 2 news plane go ahead?"

"That's a negative," replied the pilot for WABC, Channel 7.

"No sir," said NBC Nightly News.

"Negative," said ABC World News Tonight.

"Absolutely not," said CBS Evening News, showing not a hint of compassion for their New York affiliate.

WCBS, Channel 2, locked his brakes and began the $5-a-minute wait.

Chick Gallina of the NRC summed up the feeling at Three Mile Island on Thursday morning:

"It was a lot calmer than it was on Wednesday. By that time—

I don't mean to imply that it was any type of panic situation, but more assurance was felt that the reactor was stable . . . There were puffs here and there, but not a constant radioactivity, and the off-site levels had gone down. We appeared to know where the releases were coming from, and it was a matter of pumping the water off the floor into the tanks, which was ongoing. Once this occurred, the releases we were seeing dropped dramatically, and things started to get into a more recovery-oriented atmosphere, rather than the emergency affair that existed the day before.

"We still had significant problems with respect to health physics, but that was all on-site. Our concerns off-site had diminished many orders of magnitude."

William Dornsife of Pennsylvania's Bureau of Radiological Protection was at the plant that morning. As Pennsylvania's only nuclear engineer he'd been sent by his boss, Tom Gerusky, to assess the situation. Dornsife said, "I got the impression that the thing was winding down by that time—that it was essentially over."

Reilly and Dornsife sent a platoon of Geiger-counter carriers to scour the fields around TMI and to report any significant releases. "We thought it was as much a political [as a safety] move to do it," Dornsife said later. "We were now confirming there were off-site readings and because we knew we didn't have the communication system yet, our guys had to go and take their readings and stop at a pay phone and call back to the office."

While he was visiting the observation center, Dornsife noticed significant radiation readings, though they did not bother him much at the time: "I do remember very vividly seeing some radiation readings at the plant vent being relayed out to the communication center of well in excess of 1,200 millirems per hour that caused all the panic the next day. I remember one being as high as 3,000 millirems per hour at the plant vicinity."

The accident was far from over. The reactor core was finally covered and one of the main cooling pumps was circulating water to the core, but the plant was *not* in cold shutdown or even near to being there. A bubble of hydrogen gas floated at the top of the reactor pressure vessel waiting for another day to pass and for its turn to cause panic. Somewhere at the bottom of the pressure vessel lay the remains of the boron and silver control rods that had, the day before, melted away. There was no talk at the plant about a *nuclear* problem, because, of course, the control rods had all dropped into place and a chain reaction was impossible. But the only thing keeping the core from coming back to life, from going critical again, was the concentration of boron in the cooling water, absorbing stray neutrons and keeping the reactor at rest. It was highly unlikely that anything would happen to vary the boron concentration, but the operators running the machine did not even know it was something to be watched.

In almost every area, Unit 2 was down to a single working safety system in an industry that prided itself on backup systems. To keep the reactor from going critical, there was only the borated water. To keep the reactor within temperature limits, there was only a single reactor coolant pump. And that pump was beginning to wobble. Both Met Ed and the NRC had doubts about its lubrication system. And Met Ed did not tell the press that they were not sure they could start any of the other three pumps if they were needed. Finally, the only system protecting the public from exposure to lethal doses of radiation was the containment dome itself.

There was little doubt that the dome would remain intact, holding in the millions of curies of radioisotopes loose within it. But some reactors don't have containments. In the Soviet Union, they are thought to be an unnecessary luxury. Some American experts had contended that multiple safety systems made the bulky and expensive concrete containment structures needless. Yet here stood a containment—the final line of defense—filled with enough poison to kill thousands—and the utility and the regulatory agency did not yet know it. Weeks later, John Fialka, who covered TMI for the Washington *Star,* said, "Engineers have been arguing for years as to whether you need a dome. I think we settled that one, but it's a hell of a way to end an argument."

*   *   *

128

At the same time Creitz and Herbein were reassuring the press in Hershey, Dr. Ernest Sternglass was doing just the opposite in Harrisburg. Sternglass, a professor of radiology at the University of Pittsburgh, was a long-time opponent of nuclear power. His contention, first appearing in a 1970 article in *Esquire* magazine, was that low-level radiation led to significant increases in cancer, leukemia and infant mortality in populations surrounding nuclear power plants. Sternglass was forceful, articulate and used to dealing with the press. What's more, he viewed his mission as vital—the most important thing in his life.

"Nobody doubts Ernie Sternglass' credentials," said a prominent health physicist who didn't want to be named. "He just about invented the study of low-level radiation. And much of what he says is sure in the right direction, but I have a hard time believing all of his statistics."

Sternglass gave a press conference with Dr. George Wald, a retired Harvard biologist, Nobel laureate and antinuclear activist. Their concern was with the potential effects of Unit 2's puff radiation releases on infants and unborn children in the area around the plant.

Sternglass and Wald were not at all in keeping with the caricature of nuclear power opponents advanced by one writer: "vegetarians in leather jackets [who] drive their imported cars to Seabrook listening to the Grateful Dead on their Japanese tape decks amid a marijuana haze . . . bearded social democrats fearfully talking of power plants as if they were bombs."

The assembled press—a much smaller group than in Hershey but, with several hundred reporters in town, still a good turnout—heard Wald talk in general about the dangers of low-level radiation.

"Every dose is an overdose," Wald said. "There is no threshold where radiation is concerned. A little radiation does a little harm; a lot does more harm."

Sternglass had brought along some instruments and made tests near the airport that morning. "There is continuing gamma radiation coming out," Sternglass claimed, "but the readings that I obtained in the—on the way here from the airport indicate that it is not just direct radiation from the reactor which, in fact, goes down briefly when you go behind a mountain . . . So what we're dealing with is fallout, plain old bomb kind of fallout.

"The reaction of the community should be to stand up and scream. I think they should protest violently about this whole matter of deciding not to warn the public, not to let the authorities know for many hours after this very serious situation developed.

"I'm particularly concerned with the possibility of an increased risk of leukemia and cancer among the very young. The newborn and the infant in the mother's womb are ten to a hundred times more sensitive than the average adult . . . pregnant women should very seriously consider leaving."

Sternglass and Wald were clear and articulate. They made strong statements and seemed to have the professional reputations to back them up. They were news, and the local press should have thrilled to the story. But the local press didn't. The story appeared in the local papers that evening and the next morning, but local radio, the medium that had turned into the most important source of information about the accident for local citizens, was wary.

"I wouldn't go with a story like that under any circumstances," said Harvey Tate, manager of WCMB radio in Harrisburg and a former newsman. "Just think of the panic it would cause."

John Baer of WITF-TV in Hershey said, "His findings and statistics might have been as accurate as hell, but I thought putting him on the air at that time would have caused panic."

But one broadcaster did decide to use Sternglass' statements on the air. Mike Pintek, news director of WKBO and the man who broke the story of TMI the day before, decided to go with it. And, by airing Sternglass' claims, Pintek produced a good example of what Penn State professor Mark Dorfman called "a twenty-first-century threat to a nineteenth-century culture": Some listeners left the area in fright.

"At that point," Pintek said, "after hearing so many contradictory statements, I felt Dr. Sternglass was just as much a legitimate authority as anybody else was. Hell, what was I supposed to do, say not use him? So I said use him. Let's do it.

"Unfortunately, our disk jockey commented on Sternglass' statement and made it sound like there was an official order for women and children to leave the area. I told him to stop and said that all statements would come from the newsroom. People were calling in panic. I didn't know what to do, how to stop it, if indeed I should. I ordered the story stopped for a couple of hours."

Pintek was crushed. A young broadcast newsman who took his job very seriously, who liked to believe that he was performing a public service, saw his efforts fall apart under a flood of telephone calls from listeners.

"It was overwhelming," he said later. "I couldn't think. I didn't know what to do, so I went to capitol hill, and while I was there, I talked to Sandy Starobin."

Sandy Starobin was the Harrisburg correspondent for the Westinghouse stations in Pittsburgh and Philadelphia. He was looked up to by the local radio and television reporters who covered the state government. "I told him about what had happened, the story and all the calls and how I had pulled the story," Pintek said. "And all he said was 'Well, that just means you're a nice guy.' I called the station and told them to start running the story again."

No hard-news reporter wants to be known as a nice guy.

Calls were coming in to all agencies of the state government—angry calls, pleading calls, frightened calls. " 'Should I leave? I think I may be pregnant'—that's the question we got all afternoon," said a nurse who'd been on duty at one Harrisburg hospital. "We told them to stay put, but we were beginning to wonder ourselves."

Maggie Reilly at BRP was trying to handle the flood of calls into her office when the NRC called from Bethesda. "We keep having pregnant mothers in tears calling up here to find out what this Sternglass shit is," she told one of the NRC engineers. "I wish somebody would turn him off."

"Well, I've got a good one for you," the NRC man replied. "A guy —he's in Omaha now—went through on a train near Three Mile Island, and he's calling up all upset, so I had to take care of him. I told him to keep his socks on."

"Oh, okay. He'll probably get a vasectomy now," Reilly said.

PEMA was getting its share of calls from frantic citizens too. "My dog is dead! Dropped right over around eleven this morning," a local farmer claimed in his call to PEMA headquarters. "I tell you it's radiation poisoning. That plant killed my dog."

A near-frantic Department of Environmental Resources staffer said, "I'm still waiting for the call about a two-headed calf."

Paul Critchlow's main concern that morning was with publicity and media exposure—his boss wasn't getting enough of either. The morning news programs had been dominated by Met Ed and the NRC without a single Pennsylvania spokesman giving the state's side of the accident. With the situation rapidly clearing up, something had to be done to get some press while it was still available.

"There were not a lot of inquiries coming to us in the morning because Metreplitan Edison held a press conference that morning," Critchlow said. "And that really absorbed the press's attention at that point. We decided to get Scranton into the plant and spent, oh, an hour or two trying to arrange it."

There was already a congressional group scheduled to make a tour that afternoon, and Walter Creitz, president of Met Ed, suggested by phone that Scranton just go along with the group. But Scranton had other ideas.

Creitz, kept in the dark about Scranton's intentions, didn't understand. "Well, I won't be there," he said. "I'll be with the congressmen."

"That's better," Scranton replied. "I just want to go have a look."

("I didn't tell him at the time that I wanted to go into the auxiliary building, which was at that time where most of the radioactivity was coming from that was off-site," Scranton later said.)

Scranton went to the Island with a state policeman and with his assistant, Mark Knouse. Once through the tight plant security, they donned protective suits and went to the Unit 1 control room. Then Scranton said he wanted to go into the auxiliary building to see where the water was and where the radiation was escaping.

"The damn fool," one operator said, "wanted to go inside and wade through the water. We couldn't very well tell him no, but we made it clear that it was his decision."

Wearing a yellow rubber suit and an oxygen system, Scranton entered the auxiliary building alone. Neither of his companions seemed especially interested in following him into the humid, radio-active environment of the auxiliary building. Inside, there wasn't much to see. Scranton had wanted to see water. There was water, underneath plastic sheets spread on the floor to limit evaporation. And he wanted to feel radiation.

"I remember," Scranton said, "their dosimeter—not the dosi-meter, but the radiation—the Geiger counter or whatever it is showed about 3,000 millirems."

Scranton sloshed around the water in the auxiliary building for a few minutes. "I don't know what he expected to see in there," the operator said. "Water is water."

The stated reason for the Scranton visit was not to get publicity, but to assess the mood at the plant and to reassure the public that conditions were safe. Scranton's assessment of the mood of the plant was that "things seemed to be pretty calm down there, and they were. The people there seemed pretty calm." His few minutes in the auxil-iary room picked up 80 millirems on his dosimeter. But that evening, when interviewed about his visit on the MacNeil/Lehrer Report, the lieutenant governor made a point of saying, "I feel fine." The press loved it.

Just about the time Young Bill Scranton was wading in the auxiliary building at Three Mile Island, Richard Thornburgh was finally start-ing to call shots for the state's handling of the accident. While the day had begun well, the situation now was deteriorating. Thorn-burgh didn't trust Met Ed; Mayor Reid's display at the utility's press conference had made a lot of citizens realize that anger was an alternative reaction they might have rejected too quickly.

Dr. Sternglass' statements carried on WKBO had the town in an uproar. And added to the growing panic in the streets and homes of central Pennsylvania, there was evidence of some panic in the state government itself. The Secretary of Health had called for an evacuation.

Thursday was the thirteenth day Dr. Gordon MacLeod had been the Secretary of Health for the Commonwealth of Pennsylvania. Some of those thirteen days had been spent in California. Many of the remainder were occupied with parties of congratulations for the new Secretary. MacLeod had no expertise in radiological health and neither did his department. Its Bureau of Radiological Protection, under Tom Gerusky, had moved to the Department of Environmental Resources in 1971.

For the first day of the accident, MacLeod was kept informed of developments by Joseph Romano, the department's public information officer. Romano learned what was happening by watching the wire-service teletypes.

Early Thursday afternoon, MacLeod received a call from Anthony Robbins, director of the National Institute of Occupational Safety and Health. Robbins had been the health commissioner of Colorado during the Fort St. Vrain nuclear accident of 1978 and had ordered the nation's only radiological evacuation at that time.

MacLeod described their conversation:

"He said, 'Gordon, I am concerned about the events at Three Mile Island. I am concerned.'

"I said, 'Tony, from all the reports I have had, is that we have had very little radiation exposure as a result of the accident.' He said, 'Gordon, I am not concerned about that. I am concerned primarily about the shutdown process.' . . . He said he thought we were in an experimental mode. I think that was the word he used, and that we didn't know how to shut down the reactor as a result of this accident and he said he was in consultation with the Bureau of Radiological Health, which is part of the Food and Drug Administration . . . and based upon his experience and this consultation, he urged me to consider recommending that an evacuation take place."

Some time after the accident, Donald Kennedy, commissioner of the U.S. Food and Drug Administration, would say, "Tony Robbins always did have a tendency to do other people's jobs for them."

MacLeod immediately called the governor's office. In a conference call with Henderson, Gerusky and John Pierce of the lieutenant governor's staff, Macleod told them Robbins' recommendation. Robbins also asked about "the desirability of moving all pregnant women and children under the age of two out of the area." He was worried about much the same radiation effects Sternglass had mentioned that morning, and especially about the problem of thyroid cancer caused by uptake of radioactive iodine, a problem that was still expected but had yet to manifest itself. Gerusky and the others put MacLeod's fears to rest . . . for the moment.

Thornburgh was aware of the public panic, his own lack of faith in Met Ed and of MacLeod's recommendation. In the governor's view, it was time to take a stronger stand. Around 4 P.M., Thornburgh held his first press conference of the accident:

"Good afternoon. I'd like to address my initial remarks to the people of central Pennsylvania. I believe, at this point, that there is no cause for alarm, nor any reason to disrupt your daily routine, nor any reason to feel that public health has been affected by the events on Three Mile Island. This applies to pregnant women, this applies to small children and this applies to our food supplies. I realize that you are being subjected to a conflicting array of information from a wide variety of sources. So am I. I spent virtually the entire last thirty-six hours trying to separate fact from fiction about this situation. I feel that we have succeeded on the more important questions."

As usual, Thornburgh was backed on the platform by his experts. Chick Gallina represented the NRC. He could see no reason to change his earlier optimistic appraisal of the situation in response to what were generally political developments.

A great fear of engineers in the nuclear industry is a large-break loss-of-coolant accident in which, for some reason—such as an earthquake or sabotage—one of the "candy canes," the yard-thick pipes leading from the reactor-pressure vessel, breaks, allowing thousands of gallons of coolant to escape in minutes. Chances are, in a "large-break LOCA," as it is called, much of the radioactive water from the pressure vessel would end up in the normal wastewater system of the plant. To minimize risk of releasing radiation, then, the wastewater system is diverted during a radiological emergency.

At 7 A.M. on Wednesday, all the showers, toilets and drains on Three Mile Island stopped flowing into the Susquehanna River. They were diverted into storage tanks where dissolved radioactive gases could decay-off before the water was released to the river.

By Thursday the wastewater tanks were filled with 400,000 gallons of slightly radioactive, slightly polluted water. There was no more space to store the water, and the utility wasn't sure what to do with it. Nobody had ever expected an accident to last more than twenty-four hours. Nobody had ever thought of what to do when the tanks were full.

Met Ed wanted to dump the water, diluted maybe 50,000 to one with normal water from the Susquehanna which, too, was slightly polluted but not radioactive. But Gary Miller didn't know if he needed permission to dump the water.

If the water was not diluted and released by early Thursday afternoon, the full tanks would back up, releasing undiluted water into the river.

Met Ed's Richard Dubiel said, "The only alternative would have been to block the floor drains in all buildings in the plant. But doing that would jeopardize equipment, in that you could flood out the lower elevations of several buildings."

When Bill Dornsife of BRP was at the observation center that morning, he was asked about dumping the water. According to Dornsife:

"A Met Ed type asked me whether the state could give approval to discharge some water. I said, 'Well, I don't know.' It is our responsibility to approve this. And, yes, it is an industrial-waste-treatment system . . . And I said, 'Well, you know, I am not sure radiological releases are occurring over there . . .' Maggie was involved in it. My understanding is that Maggie said, 'Yeah, it is all right with us but you have to get NRC approval too.'"

Dick Dubiel of Met Ed called Maggie Reilly. "He indicated that if we don't get permission to dump it that the sumps are going to run over and it is going to go out as an uncontrolled discharge through a storm sewer anyway," Reilly remembered. "So you know, you don't have a whole lot of choices there, and it was less than MPC [Maximum Permissible Concentration], so I said, 'No, it doesn't give me a heartache.' ''

Dubiel even checked with Reilly and Dornsife's boss, Tom Gerusky, who approved the release as long as it was kept below MPC and communities downstream were notified before the release took place.

Dubiel also checked with NRC's Chick Gallina, who said, "As far as I am concerned, if it is within NRC limits, it can be released." The trouble was, he later explained, "that it was a gas dissolved in a liquid, and we have release limits for liquids, and we have release limits for gases. They asked, 'How do you handle this?' assuming that when you release all the gas contained in the liquid, it will be released to the atmosphere . . . I said, 'You check . . . with headquarters and make sure you get the okay . . .'"

Edson Case at the IRC in Bethesda remembered being told about a 2:30 P.M. call from Met Ed, "checking with us" about releasing the water. Case thought the dumping was a "a dumb thing to do, because it's a political question, not a technical question . . . and my gut said you're going to get a ruckus about it . . ."

Lee Gossick approved the release, which began at 2:45 P.M., just as the congressional delegation was making its tour of the stricken nuclear plant.

Late that afternoon, not knowing anything that had transpired before, NRC Chairman Joseph Hendrie and NRC Commissioner John Ahearne called IRC to find out what they could about the release. They talked to Edson Case and John Davis, both experts in Bethesda who had no idea Lee Gossick had already approved the release. They were just as confused as the NRC Chairman and, through their ignorance of the situation, made things seem far worse than it really was.

HENDRIE: What's going on with this dump down at Three Mile into the Susquehanna . . . I just got a report they'd released 400,000 gallons of slightly contaminated water into the river.

DAVIS: Mr. Chairman, we aren't certain of that. We had a report at 2:30

137

this afternoon that they were releasing some controlled release into the river at about 10 to the minus 3, 10 to the minus 4 [concentration of xenon per liter of water].

HENDRIE: Okay. That's consistent with the report that's come through. Now, where did that come from?

DAVIS: The licensee. It comes from the licensee. Now, it's our understanding that it is a controlled release . . .

AHEARNE: Are we letting them do it?

HENDRIE: They have done it, or what is the situation?

DAVIS: It was my understanding they were in the process of controlled release. Whether the 400,000 gallons have gone out, I don't know; we'll have to check.

HENDRIE: I thought they weren't going to do things like that without letting us know?

CASE: Well, they let—as I understand it—they let us know they were dumping. They maintained, I gather, that it was in the licensed limits.

AHEARNE: Did you perchance question them and tell them not to do it until they ask us a little bit about it?

CASE: That was one of the messages that went back at one point.

HENDRIE: Now, if Three Mile were operating normally, then the licensee might find it within his license that he can go ahead and make this release. That would be all right. In the circumstances, the impression everybody will have is that he is dumping the contaminated water into the river.

CASE: Bad PR, agreed. Why don't we just call them right now and tell them to stop, if he hasn't stopped it?

HENDRIE: I think something like that would be more useful if we had started it a little earlier. This may be a separate set of water someplace, and if we can all shake hands that "Oh, this is fine, you walk through a tank, this is another set of stuff; there is a trace of stuff in it, but it's well within limits," and so on, okay. But I don't find it very happy to have him just crank a valve and running this stuff into the river. You know, supposedly we've got a team down there that's keeping track of things and I'm going around telling congressmen we have good, close communications and that we and the state people and everybody else are working closely together so that we all know what is going on, all agreed on the steps. And I don't find that compatible with him just deciding "What the hell, I'll dump 100,000 gallons," even if the level is, in fact, minimal . . . He's not running a normal, everyday configuration down there, for God's sake.

CASE: Well, word has gone back through our chain for the licensee to stop.

HENDRIE: Okay. Well, maybe it will turn out he hadn't opened the valve and was checking. I don't know. So do you have any reflection down from the site about what the state people are going to—

CASE: The state was, as I understand, aware of it.

HENDRIE: Like us, they were just told he was dumping it?

CASE: Yes.

HENDRIE: Jesus Christ! I bet they're calling the governor. Those goddamned fools down there are dumping their stuff in the river, they can't think what to do with it. Would you please get hold of the state people and find out what's going on?

Hendrie's order to stop the release went down through the chain of command from H Street to Bethesda to King of Prussia to Chick Gallina at Three Mile Island to Gary Miller, who reluctantly stopped the discharge after 40,000 gallons of wastewater had flowed into the river. Miller couldn't understand why the NRC, which had already approved the release, were now telling him to stop it. But Miller knew it would only be a temporary halt—there was no alternative but to dilute and dump. At the same time he ordered the release stopped, Miller also ended all showers and flushing of toilets on Three Mile Island.

George Smith had been manning the phones at Region One headquarters in King of Prussia when the call from Bethesda came in. Smith tried to explain to the IRC that the controlled release had to continue or "it will go over the side, because it's overflowing." Bethesda was of no mind to listen to advice from George Smith. "I just got hell for asking questions," he said.

Karl Abraham explained the political nature of the release to Chick Gallina in Harrisburg. "Karl told me that they are releasing industrial wastewater," Gallina recalled later, "and that it may become a problem because the word 'release' was becoming a dirty word, and that anything to be released was something to be careful of because the press gets excited and politics may take over . . ."

Politics *had* taken over. Edson Case had been right that the problem was more political than technical. Shortly after Hendrie ordered the release stopped, it became evident in the IRC that it would have to be resumed within a few more hours. There was just no choice. And, since it was the NRC that had ordered the dumping stopped, someone had to order it started again. In the early evening, Karl Abraham and his boss, Joe Fouchard, began to figure a way to ensure that the NRC didn't have to take responsibility for restarting the release.

In an office in the west wing—the executive wing—of the state capitol in Harrisburg, Karl Abraham of the NRC had set up shop on Thursday morning. Abraham was the public information officer for NRC's Region One. His job was to tell the press and the public what the NRC was doing to fix the problems of Unit 2. On Thursday Abraham still had to do it all alone. He was the only NRC public information officer in Pennsylvania, one of eight in the entire nation.

Joel Shurkin of the Philadelphia *Inquirer* had covered the space program with Abraham when the latter was a science reporter for the competing Philadelphia *Bulletin.* "Karl was the most organized reporter I ever saw," Shurkin said later. "He had this sort of portable desk he kept at the space center in Houston. He'd roll it out and there would be a typewriter, reference books, slide rule, everything he needed. It was great. We used to be good friends, you know."

Abraham "used to be good friends" with a lot of reporters, but that relationship changed when he went to the NRC. Richard Roberts of the Harrisburg *Patriot* recalled: "I'd had some dealings with Karl Abraham and he was not cooperative—difficult, personally, to deal with . . . He will use any excuse he can not to be helpful."

Others of the press were not as polite as Roberts. Jerry Ackerman of the Boston *Globe* said, "When I've dealt with him in the last three years, he's always adopted a heels-in-ground posture. He was no different in Harrisburg than he was anywhere else—an arrogant son of a bitch."

And Curtis Wilkie, also of the *Globe:* "Abraham was not very communicative. He tended to dismiss me on the grounds that I didn't know enough about the subject to ask intelligent questions, which I thought was a hell of an attitude for a flack to have. He was condescending, always in a hurry, running around like a chicken with his head cut off. I said the hell with this guy."

Even Abraham's co-workers, the other seven NRC public information officers in the United States, had trouble getting along with him. "We call him 'the prince,' " said one of them, "because he thinks he is our ruler."

By early Thursday night, the only press inquiry about the dumping came from Dick Lyons of the *New York Times.* Lyons asked Abraham to check out the story he'd heard from his editors in New York about a release of radioactive water. Abraham denied the story. In Washington and Bethesda, the line being given the press was "We're checking with the licensee."

Governor Thornburgh first heard about the dumping and that the NRC had ordered it stopped sometime after 6 P.M. He asked his technical people for more information and soon came to believe that the NRC had stopped the release so he would have time to decide whether it should continue.

When Hendrie heard from Fouchard that "we're getting word here that the governor himself wants to have a finger in this pie," he responded, "Fine, let the governor decide whether we are going to dump the crappers in the river."

Gerusky and Reilly knew that the release had to be continued and told that to Thornburgh. Maggie Reilly saw through the political nature of the problem and the difficult time both Thornburgh and the NRC were having in dealing with it. "I thought, boy, if it were a *real* problem, what do you do then?" she recalled.

Dave Milne of Critchlow's staff began to write a press release placing responsibility for resuming the release squarely on the Commonwealth of Pennsylvania. He showed the release to Karl Abraham, who further honed it to lay the blame squarely on Thornburgh and DER:

"The Department of Environmental Resources tonight authorized Three Mile Island Nuclear Power Plant to discharge into the

Susquehanna River 400,000 gallons of wastewater that has become slightly contaminated with xenon, a short-lived radioactive gas . . ."

Milne and Critchlow began to have second thoughts about whether this time it was really the state's legal responsibility to approve resumption of the dumping. Abraham said the NRC had just stopped the dumping to be "polite" and give the governor more time for making a decision.

"We had a question whether the governor had authority which superseded the NRC's to permit or approve the dumping," Critchlow recalled. "After much pulling back and forth, we got Karl to acknowledge that no, he did not."

Dave Milne threw away the draft press release, thinking, "Why should I do it? I mean, why should we take responsibility for this when in fact we don't have any authority over it?"

Sometime after 10 P.M., there was a confrontation between Paul Critchlow and Karl Abraham. They sat in an office on the second floor of the state capitol. The furniture was heavy and deeply carved oak. Near the fifteen-foot ceiling, the room was lined with portraits of former governors of Pennsylvania, looking down on the men.

Critchlow had decided to protect his boss. "I'm not going to let the governor be associated with this stigma if he doesn't have to be," he said.

Searching for any way out of what he was sure would be an enormous problem with the press, Abraham replied, "Well, you *do* have a lieutenant governor."

Seated next to Abraham at the table was Mark Knouse, Scranton's executive assistant. "I basically said nothing," Knouse recalled, "just sat back there and seethed quietly. Abraham didn't know who I was . . . Paul looked at me and I think I made some expression."

Like Met Ed had before it, the NRC had just used up its supply of credibility—credibility it would need later on in the accident. Critchlow didn't trust Abraham or his agency. The next time the NRC would look to Thornburgh for quick action, it would be to find the governor acting in slow motion, wary of being taken advantage of again.

Dave Milne's second-draft press release was finished just after midnight on Thursday. By the time it was available in the capitol newsroom, nearly all of the reporters were gone. Most of them had never known that the dumping took place. In eight hours the press would be back, up in arms about being kept in the dark for nearly ten hours about the release of water.

Karl Abraham moved his office out of the state capitol the next morning.

Prior to Thursday's dumping incident, the important story at Three Mile Island had been the reactor, its condition and the possible threat to public health presented by the closest brush with a meltdown ever experienced by the U.S. nuclear industry. But with the dumping incident and its politics and in-fighting, the whole nature of the accident changed.

After Thursday night, there would be further concern about radiation and public health. Indeed, the greatest concern, the greatest fear among both experts and the public, was yet to come. On Wednesday the system had survived, barely, all the incompetence its operators, engineers and executives could muster. It would be weeks before those men would even guess Wednesday's danger, and they had still to face the fears of Friday, Saturday and Sunday, when the danger would seem greatest but really wasn't.

At this point, the people—mostly men—of the NRC, of Met Ed and of the government of Pennsylvania, rather than the plumbing, become the new story of Three Mile Island. Whereas the machinery had gone mutant on Wednesday, compounding errors, compounding failures, now the politicians and technocrats would take their turn. For the next few days, while the reactor worked to correct its own problems, the public and press would be violated by fighting between agencies, within agencies, by those trying to take political advantage or trying to save their political necks, by a new school of scientific thought that says fifty experts following the same false premise can't be wrong, and by a few key men who just couldn't bring themselves to tell the truth.

15

A Problem of Information

Thursday was the most frustrating day for the press at Three Mile Island. They were all there—hundreds of reporters, photographers, cameramen and producers—with very little to say. The story wasn't very visual. It wasn't well-suited to television. The cooling towers, looming 300 feet into the air, made an interesting picture, but one that didn't change to suit the news event. They didn't move, didn't grow, didn't do anything. The story was stagnant. Only Jimmy Breslin of the New York *Daily News* was able to see "evil steam, dripping like candle wax down the cooling towers."

Bryce Nelson of the Los Angeles *Times* said, "What TMI most reminded me of was Attica, where there was a story going on inside these walls and you really couldn't see any evidence. Certainly in the first few days, Met Ed made the attempt to be open. At Attica they wouldn't even have press conferences."

It was obvious to the press that, like Attica, the story was all on the inside. And getting to the inside was becoming more and more of a problem. To Rod Nordland of the Philadelphia *Inquirer,* though, it was more of a challenge. "Our investigative team had been sent in primarily to interview workers at the plant," Nordland said, "to find out the background of this accident. We tried to interview workers coming off-shift at the plant gates, at the observation center, anywhere we could get to them. They wouldn't and, in some cases, couldn't talk. The company wanted its employees to be quiet and gave orders to that effect, so it was virtually impossible for us to do any interviews around the plant or in sight of the other workers."

Most of the Met Ed employees were happy not to talk to the press. Often, they thought the reporters were making more of the accident than was proper. And many of the hourly workers were afraid for their jobs.

A clerical worker named Mary called her mother that Thursday to say that everything was all right. The next week, the weekly paper in Mary's mother's town—the sort of paper that announces when grandchildren come to visit and pets die—included a sentence: ". . . and Mary called to say everything was fine at Three Mile Island." "You wouldn't believe the trouble I got into for that story," Mary said later.

Nordland wanted to contact the workers at their homes if possible, but Met Ed would not release a list of employees. So, reporters for the Philadelphia *Inquirer* took turns writing down the license numbers of all cars entering or leaving the nuclear plant. Each number was put on a notecard, and the hundreds of notecards were put in a shoebox.

It was just the kind of tedious, monotonous work that reporters hate to do, but Nordland used remarkable powers of persuasion to motivate his workers. He had twenty-eight reporters, the *Inquirer*'s entire metro reporting staff, and was still asking for more. By the weekend the *Inquirer* would even begin planning to send the city staff to Harrisburg as well, leaving the news of Philadelphia in the hands of the sports department.

A day of noting license numbers, and Nordland went off to tap the Bureau of Motor Vehicles computer, returning with names and addresses for each car that had traveled to the Island. That night the *Inquirer* sent three hundred Mailgrams to the names on Nordland's computer printout. The Mailgrams offered anonymity to workers who were willing to talk about previous problems at the plant. They listed a number to call. And they got results. The fact that the Mailgrams were also received by Met Ed management and at least one reporter for the Associated Press did not dim the *Inquirer*'s triumph. They soon knew the many previous problems of Three Mile Island's Unit 2—the worker dissatisfaction,

145

massive mandatory overtime and safety shortcomings of the plant.

But while the *Inquirer,* with its army of reporters, had the man-power to track down a hundred Met Ed workers, many other report-ers had to work alone. They had to rely on more conventional methods and traditional sources of information. And most of those sources were either not available or were just incapable of answering technical questions.

The Commonwealth of Pennsylvania was a hopeless source of information. Theoretically Paul Critchlow and his staff in the gover-nor's office were responsible for any statements by Pennsylvania. Critchlow had decided to take this course, shutting up the state's many involved departments, after John Comey, PEMA's only public information officer, had gotten carried away with being a media star on Wednesday. Comey speculated on any subject and at any length for the national press who were listening to him and his agency for the first time in both their careers. On Wednesday morning he told Mary Bradley of the Harrisburg *Evening News* that the reactor had "failed fuel" without either of them knowing what that meant. The statement was true, of course, but it lacked restraint. And Thorn-burgh demanded restraint, control, an organized and dignified ap-pearance. That was Paul Critchlow.

Yet most reporters found Critchlow to be no source at all.

"I have a bone to pick with Paul Critchlow," said Mary Bradley. "We were told the governor's office would relay all information from the state, but his people were always busy. They wouldn't return calls. They wouldn't even take messages! It was very difficult to get anything out of the office at all. I'm not sure we did."

Critchlow and his staff were unable to adequately serve the needs of the press because they were forced into what was to them more important service—acting as the eyes and ears of Governor Thornburgh.

"It became my responsibility," Critchlow said, "to try to get the most solid, reliable facts and data that I could for the governor so he could make his decisions . . . And it was a very intense process in the governor's office itself, which was sort of a command post to cross-check and call up the range of sources, get hold of all the different people, experts that you needed, to arrive at the point where you were ready to go out and say something."

Eventually the state's effort at gathering information extended to the point where a rumor-control telephone center was set up, not so much to control or put down rumors, but to tell the governor what rumors were floating around the state. The public was not told the real purpose of the phone center.

But at least the governor gave press conferences to relay much of what he had learned. Many other likely sources did not even do that. Babcock & Wilcox, builder of the Unit 2 reactor, was particularly silent. John Emschweiller of *The Wall Street Journal* was typical of most reporters in his lack of success with the reactor builder. "The most interesting thing I got out of B&W," Emschweiller said, "was that they'd just hired a new public relations man named P.R. Miracle."

Having been stung more than once by the press, Met Ed was becoming increasingly less cooperative. George Troffer seemed to be the only source of consistent and reliable information available within the utility, and a number of reporters were calling him several times a day for updates. Most of these reporters, including Ward of CBS and Casey Bukro of the Chicago *Tribune,* thought they had some exclusive "inside source."

Met Ed had been used to dealing with a local press that didn't really care or pay too close attention to the nuclear industry. "I never had any trouble with them," recalled Jim Moyer of WHP radio. "Of course, the stories were always about a power failure or a rate increase."

And the press had little more success dealing with the NRC. The experience of *Newsday*'s Stuart Diamond was typical: "When I asked the NRC, they said the answer is in the public documents room, which is only one million pages of material open nine to five, and I asked the question six at night."

Peter Stoler of *Time* badgered NRC's Joe Fouchard to allow a pool reporter and photographer into the control room, but Fouchard told him Met Ed president Walter Creitz threatened to get a restraining order if Fouchard let the press in.

147

To many of the reporters at TMI, and especially the science writers, the public-information effort of NASA during the space program was the prime example of how the job *should* be done. NASA provided experts at any time to speak honestly on any subject. Their well-informed public-information officers virtually did the press's job for them and, indeed, a number of science writers were said to cover the space race from cocktail parties.

But it hadn't always been that way. In the early 1960s, the space agency had been very secretive—often not even telling the press that an unmanned launch was about to take place. The more diligent reporters staked out Cape Canaveral, watching from nearby beaches. NASA's secretive press policy changed, though, after a 1967 launching-pad fire killed three astronauts in their Apollo spacecraft. The space agency said the three died instantly, only to have John Noble Willford of the *New York Times* produce a recording of the spacecraft communication channel with the astronauts clearly calling for help as they burned to death in agony. That day in 1967 taught NASA that, in press relations, honesty *is* the best policy.

Though Dr. Sternglass appeared in Harrisburg on Thursday, there was very little organized antinuclear activity near Three Mile Island. It should have been a prime opportunity for antinuclear organizations to air their views, but reporters had mainly to rely on telephone interviews with representatives of the Union of Concerned Scientists, Critical Mass or the Abalone Alliance. Chris Sayre of TMI Alert was alone among local voices. Most of the rest had already left town.

"I was not so surprised that the antinuclear groups were not there," said Cristine Russell of the Washington *Star*. "This was their greatest fear. Why would they purposely walk into their greatest fear?"

But if antinuclear spokesmen were not in Harrisburg and Middletown, plenty of opinionated citizens were. News coverage on Thursday was dominated by local citizens reacting to the crisis. "We couldn't keep the locals off the camera," Richard Wagner of CBS recalled, "whether it was expressing outrage or wanting to say there was nothing the matter."

Many of the stoic central Pennsylvanians were still saying that it didn't matter—that the danger was past, nonexistent or worth the risk for reliable energy.

"Newspapers like to print things that scare people," said Harry Drake, a carpentry instructor. "We need the power. I think it's worth the chance so I can have an air conditioner."

Dairy farmer Paul Lytle III wasn't worried: "If it was something you could see, maybe you'd worry more about it. This doesn't bother me at all."

"It's just like the river," said Mary Russ, who'd run the Blue House Bar for fifty-three years. "I've been through five floods at my cottage on the river across from the plant and I've always gone back."

But enough residents of Middletown, Goldsboro and Harrisburg were worried about what was happening this time—especially the farmers, who began to look at the plant as the cause of many of their problems.

"We lost two three-hundred-pound steers, one in December and one in January," claimed Mrs. Ruth Kaufmann of Fairview Township. "Then we lost the cats—four of them died in one week in December. The electric company keeps answering us there is no leakage . . ."

Brittany spaniel breeder Paul Molowka of nearby Zion's View said, "Last year I lost twelve dogs. They died of cancer. That never used to happen. The wild game started to disappear around here and I must have lost seventy to eighty head of cattle and the vet wasn't able to tell me why they died."

Most especially, pregnant women and parents of young children were afraid.

"I've never thought about it," said Janet Reisig of Middletown, who was seven months pregnant. "We don't really know that much about it. People don't know what radiation is. You just have to take the government's word that it's safe. But if I thought there would be damage, yes, I'd leave."

And some had already left. The Bainbridge Elementary School, closest school to TMI, showed only 20 percent attendance on Thursday. Parents were taking their children out of school and sending them to stay with friends or relatives out of the area.

But Alma Weiss of Limerick was staying right where she was. "I'm really a Christian," she said. "I don't live in fear because I'm walking with the Lord. I accept what he has."

"But Alma, this is man-made," responded her neighbor Charles Frisco. "I'm shook up. I'd like to have a Geiger counter and test that area myself and see how far that stuff spread. Then I'd know how much they're hushing up."

Much of Thursday afternoon had been spent trying to obtain a sample of primary coolant from the reactor containment. The hot lab, where such samples are normally taken through long drain lines, was very "hot" indeed from radioactive coolant that filled sample pipes to the room. While the NRC was searching the Oak Ridge National Laboratory and MIT for a robot to take the sample, Met Ed just went ahead and took a sample. Working as quickly as possible, the utility's auxiliary operators ran in and out of the hot lab dressed in protective suits. One man would run in to open the valve, the next man would run in to take the sample, and a third man would run in to close the valve. Careful teamwork left none of the workers overexposed on Thursday, and the NRC finally had its primary coolant sample to be analyzed in their mobile laboratory.

The news was bad—far worse than anyone had expected. Chick Gallina's appraisal that the trouble was over had been far from accurate. "I was shocked to hear the magnitude of radiation coming from the sample," Jim Higgins said. "It was much higher than I had ever expected or had ever heard of before . . . that was really the first indication that I had [that] . . . there was as much fuel damage as there really was. That really brought it home to me for the first time how serious it was."

Jim Higgins tried to call the governor. Instead he reached Paul Critchlow.

Higgins told Critchlow that the high radiation levels and tremendous amount of fuel damage—certainly more than half the fuel in the reactor had failed—suggested a greater problem with radiation leaks.

Critchlow tried to confirm the information through Karl Abraham, who called Washington and told Critchlow, "I can confirm there is more fuel damage. It does mean more long-term problems, but no, it does not mean that we will have sustained low-level radiation leaks."

Once again Critchlow had to decide whom to believe—this time between two different sources in the NRC. But neither interpretation was available to Critchlow's boss. At that moment Governor Dick Thornburgh was having troubles of his own in Hershey.

Just about the time Jim Higgins was calling Critchlow with bad news from the primary-coolant sample, Thornburgh was walking into the studios of WITF-TV on the corner of Chocolate Avenue and Cocoa Lane in Hershey. Thornburgh was eager for media exposure and found his chance to present the state's, and his, good work to the people of Pennsylvania in a statewide broadcast over public television that night.

Thornburgh appeared with moderator John Baer, a host and producer at WITF and former reporter for the Harrisburg *Patriot.* The only other guest was Chauncey Kepford, a long-time antinuclear activist who had "intervened"—represented antinuclear interests—in the licensing of both Units 1 and 2. Kepford already knew the results of the primary-coolant analysis from a contact in the NRC. Thornburgh did not.

BAER: Governor Thornburgh, you, I'm certain, are being kept up-to-the-minute on conditions at the plant and around the plant. Can you tell us now what the current situation is?

THORNBURGH: At the present time we are advised by NRC and environmental and health officials of this state that there is no off-site concern for residents. That important aspect appears to be stabilized . . .

BAER: Dr. Kepford, there have been reports all day about the levels being monitored at the plant. Are those reports to your satisfaction?

KEPFORD: With regard to the reports inside the plant, John, I'm not sure. I've heard a number of values, the highest value I have heard comes from a confidential informer inside the NRC itself and that is a very high number. It turns out to be 10,000 rems per hour. This radiation level would be indicative of a serious amount of fuel failure. This is a new plant . . . there should have been no fuel-rod leaks, yet clearly there has been a massive breakdown of that initial barrier, the fuel cladding. Now, whether or not there is fuel melting, it seems likely there has been . . .

BAER: Governor, given the fact of the conflicts, are you personally satisfied that the numbers that the state is getting now are accurate?

THORNBURGH: . . . I'm satisfied that the balance of the information that has been accumulated by the NRC and by the Department of Energy, our environmental and health people, gives us a fair reading of what the threat to any inhabitants—any off-site individuals—would be, and we are satisfied that situation has stabilized at the present time . . . because the plant is still leaking, there is always the possibility of a further deterioration . . .

BAER: Governor, there are, as you know, a series of recommendations that the state has the power to make. I think the most simple among them is just that area residents close their windows. Have any of those recommendations been suggested so far?

THORNBURGH: At the present time, the advice that we have received from the NRC, from our own departments of Health and Environmental Resources, is that the off-site hazard to individuals has been reduced to a level of insignificance and the on-site hazards are being compensated for by the individuals who are engaged in shutting down the operation and in dealing with the disposition of residue by their own internal protective devices.

BAER: Huh?

Seconds after the end of the show Paul Critchlow called Baer. He was upset at the poor way his boss had been presented. But it was Thornburgh's own fault. "It became clear during the broadcast," Baer said, "that, although he was saying he had a handle on things, he did not. I thought we had cemented our relationship with the new administration by doing a statewide special on the transition of government. So we knew the people in his press office well. I think that's how we got the governor that night. But we've never gotten him since."

*　　*　　*

It was just after midnight as Thornburgh pulled in to the side gate at the Governor's Mansion. At about that time, mimeograph copies of Dave Milne's second press release about the wastewater dump were being distributed to empty desks and mailboxes in the capitol newsroom—time bombs, waiting for morning. Kevin Molloy was getting ready for another night on his cot. Rod Nordland was hidden near Three Mile Island with his computer-controlled radio scanner, frantically trying to find the NRC's internal radio frequency for a little snooping. Mayor Reid, always afraid of being burned to death in his sleep, was testing his smoke detectors. Most of the press had gone to bed or to a bar. Harrisburg, Middletown and Goldsboro were quiet, with only occasional helicopters flying over from floodlit Three Mile Island.

In nine more hours, all hell would break loose in central Pennsylvania. But now, all was calm and a typewritten notice was posted on a bulletin board in the capitol newsroom:

To all out-of-town reporters: If you go out tonight onto the streets of Harrisburg and see that they are dark and empty, this is not—repeat not—because of the accident at Three Mile Island. That's the way they always are.

Thank you.

16
Black Friday

Friday began a little cloudy with occasional patches of rain, but still mild for Pennsylvania on the second to last day in March. Most of the major characters in the developing story of the accident at Three Mile Island were at least rested.

Colonel Oran K. Henderson felt in reasonable command of the situation on Friday morning. The accident had tested his forces and not found them wanting. PEMA had served its function well, informing local communities of the current situation at the plant and helping to finally prepare the evacuation plans that only a crisis of this magnitude could have stimulated. Henderson was right in the middle of the action, too. He was involved in all of the state's high-level discussions, was briefed by the NRC along with the governor and had gladly given many interviews to the press, seated at his desk in the fortified basement of the Transportation Building, a photograph of an atomic explosion mushroom cloud behind his head.

Perking coffee, sizzling bacon and the sound of WSBA radio filled the Pennsylvania Dutch kitchens that morning.

"Big danger is over at TMI . . . and a poll shows most persons feel they are safe from serious contamination . . . These stories headline WSBA news at eight o'clock . . . If you listen to Nuclear Regulatory Commission investigator Charles Gallina, the big radiation threat is over . . . Nuclear experts say, however, that highly radioactive water is still beaming radiation and it might have to be buried in lead or evaporated . . . Lieutenant Governor William Scranton toured the plant yesterday, along with several senators and WSBA-land congressmen . . . Senator Gary Hart says no judgment can be made on nuclear power as a result of Wednesday's accident. He was given a tour yesterday, too . . . An Associated Press/NBC News survey shows that most folks living around TMI are not too concerned. Most feel they are safe from serious contamination. Half of those questioned view the mishap as a crisis, while 37 percent do not . . . What is the radiation level at TMI? We'll have the answer coming up. WSBA newstime 8:07 . . ."

That morning, Jim Floyd had to make an important decision as the shift supervisor of Unit 2. The makeup tank—a tank of borated water used by the high-pressure injection system to add water to the reactor core—was suffering some gas pains. Radioactive gases had built up over the water in the tank, forcing the water level down. "I was losing water from my emergency core cooling source," Floyd said. "I might have needed that water later."

The gas pressure had to be relieved somehow, either to the open air or into the waste-gas decay tank. Given the high radioactivity of the gas involved—mainly radioactive xenon—and the political situation, venting into the waste-gas decay tank was the only reasonable option. But there were leaks in the transfer line from the makeup tank, and Floyd knew it. There was further concern that once Floyd opened the valve on the makeup tank, it might not close.

"If my radiation levels went high, I could reclose the valve and protect the public," Floyd recalled, "but if it failed to close, if Murphy was at his best and decided to bite me right when I didn't need to be bit, then I would have needed the public moved, and on that assumption I wanted to know what Civil Defense preparedness was to move people, and so I called them."

Floyd sent a helicopter up above the plant to be ready to measure radiation released by the venting procedure, then tried to call PEMA before the release, to let them know what he was doing.

Carl Keuhn took the call. Floyd did not seem calm. According to Colonel Henderson, "I think Carl Keuhn's words to me—I would not swear to them—were, 'Hey, this guy is going ape!' I gathered it was a very emotional, frightening report."

A Bell Jet Ranger helicopter hovered in the still air above the nuclear plant as Floyd cracked the valve to vent the radioactive gas. Six hundred feet above the auxiliary building, the 300-foot cooling towers jutted up at the aircraft like mountains. Facing southeast—into the wind—the pilot and the health physicist seated with him could see the whole plant beneath them. A little steam drifted from the cooling towers of Unit 2. Unit 1 was cold. To the right was the village of Goldsboro, just across the river. Yellow buses were picking up children for another day of school. A couple of miles behind the hovering machine, Middletown was stretching to face the third day of the accident. Seven miles further on—downwind—were the 60,000 people of Harrisburg.

As the valve opened, there was no visible change in the plant, no obvious release of radiation. Perhaps just a few more wisps of steam, that's all. Yet the sensitive Geiger counter, mounted in the cockpit, began to measure increasing amounts of radiation. From the 3 to 5 millirems per hour it had read when they settled into place above the towers, the pilot and health physicist saw the needle start to rise— 100 millirems . . . 500 millirems. Within 90 seconds, it peaked at 1,200 millirems per hour. Four hours in the plume would exceed the EPA's yearly radiation allowance for workers, if the figure did not begin to drop. Minutes passed as the turbine-powered helicopter strained to remain in just the same position, jockeying in the air currents that flowed above and around the immense cooling towers. The needle stayed pegged at 1,200 millirems for minutes and then, slowly, it began to drop. Through 1,000, 800, 500—back to just a little above where it had started. The valve had closed. The helicopter turned left, back to the wheat field that had become its home.

A second call to PEMA was logged at 8:40 A.M.: "Call from Three Mile Island Control Room—release in progress began at 0832. A site emergency has been declared. Reading 14 millirems at site fence. Six hundred feet—1,200 millirems per hour over facility."

". . . from what we had, and the way it was related to me," Henderson said, "I was about ninety percent certain that we were going to execute an evacuation."

156

Henderson barked orders to his subordinates in the PEMA communication center, ordering activation of Civil Defense units throughout the area. The orange and gray PEMA headquarters soon had every one of its 127 telephone lines in use. Colonel Oran K. Henderson was ready for his first action since Vietnam.

But Tom Gerusky and Maggie Reilly of the Bureau of Radiological Protection didn't share Henderson's concern. "We knew that levels were detected off-site, that the helicopter was up there tracking the plume, as it had been doing," said Gerusky. "In talking with the plant and NRC and our people and DOE, there just didn't seem to be anything that would have caused us to panic . . . If the levels did not start to go down, we would have been concerned. Levels started to go down immediately."

The night had been peaceful in the NRC's Incident Response Center in Bethesda. The last major problem—the wastewater release—had been a political one and of little concern to the technical experts in Bethesda. They'd spent much of the night trying to divine the level of core damage from results of the primary-coolant tests.

By 9 A.M., the main room of the IRC was crowded with men, and phones rang constantly. Harold Denton, head of NRC's Operating Reactor Division, was meeting with his lieutenants in the glass-walled executive management team room.

As far as the NRC was concerned, the reactor was stable. But the information gleaned from the previous night's primary-coolant sample suggested much more production of radioactive gases in the system than they had originally suspected. So a new fear developed—that the waste-gas decay tank would be completely filled, opening an automatic overpressure relief valve and releasing radioactive xenon and who-knows-what-else into the air above Three Mile Island. The question of the hour was: How much radiation would be released if the tank couldn't be vented back to the reactor containment?

This was a completely different problem from the one Jim Floyd had faced that morning. Floyd was worried about shifting gas *to* the waste-gas decay tank. The IRC concern was over the amount of gas *already in* the tank, a concern that was not justified. The tank was not even close to being filled.

Lake Barrett was an NRC engineer who had been calculating radiation doses from the accident since Wednesday afternoon. Shortly before 9 A.M., Barrett turned his attention to how much radiation would be released if the waste-gas decay tank was overfilled and had to be vented. The answer was 63 curies per second—a very significant amount of radiation. John Davis, a member of Denton's executive management team, saw Barrett's calculations and their implications. He asked the engineer to explain his results to Denton and the others.

While Davis and Barrett were consulting in the outer room, Harold Denton was learning of Jim Floyd's 1,200-millirem release from Karl Abraham, who had called from Harrisburg. Abraham wanted to go beyond his function as a public-information officer and be a part of the NRC decision-making apparatus, so he called Bethesda when he heard about the release from PEMA. He told Denton and the others that there had been a 1,200-millirem release, but when questioned about where around the plant that high number had been measured, he couldn't say. The main possibilities were that the reading was taken on-site, at the plant, in which case it was not as bad as some readings of Wednesday and Thursday, or that it could have been measured off-site, across the river in Goldsboro or at the observation center, in which case it would be above Environmental Protection Agency safety levels and an evacuation would be required. Not knowing which it was, on-site or off-site, someone at Bethesda arbitrarily decided "It must be off-site."

The actual reading mentioned by Abraham was measured 600 feet above the plant and would have diminished by at least a factor of ten before it reached any populated area, but Bethesda decided to believe that the reading had been taken "off-site," that it was an indication of serious danger to the local population. There was no reason why the NRC decision-makers should have assumed the reading to have been off-site, nor any reason why greater effort was not made to check out the location of the 1,200-millirem release.

Davis and Barrett walked into the EMT conference room just after Abraham's call. At that point the top experts of the NRC knew nothing but that there had been a 1,200-millirem release from *some* part of Unit 2 measured *somewhere* around the plant. Barrett explained his calculations to his superiors of the EMT. The calculations were completely unrelated to what was happening at that moment on Three Mile Island, but Denton and the others were panicking and not inclined to believe that. Denton asked Barrett what would be the "off-site consequences" in millirems of such a 63-curie-per-second release. Barrett did a quick calculation in his head and replied, "1,200 millirems per hour."

"Oh, my God!" said Edson Case.

Barrett had estimated 1,200 millirems per hour from the waste-gas decay tank and they had just heard of a 1,200-millirem-per-hour reading from Three Mile Island. The tank must be leaking as Barrett had feared!

"My perception was that that had a very profound impact on the whole center," Barrett recalled, "that we had shifted from a sort of lack of information on things and nothing really firm to 'Well, here is a real piece of meaty information that has significance to it.' I believe it took a hypothetical situation and rather carved it in stone and set it on a mountain with a burning bush behind it . . . People immediately started talking about evacuation."

These two pieces of information came together in Bethesda at a time when the NRC was suffering profound confusion about what direction to go in ending the crisis. "There were increasing concerns on all sides," Denton said. "It was not just the releases, but by Friday morning the whole picture was one of uncertainty, not being sure of the actual status of the case and I sort of saw this puff release of radiation as being the last straw. It kind of destroyed my confidence that we really knew what was going on up there . . ."

Denton turned to Barrett and asked how far from the plant should citizens be evacuated? Barrett had no experience in emergency planning and said as much. He couldn't make a recommendation. In panic, Denton asked him again.

Barrett, who had no experience with emergency planning, no

detailed knowledge of the health effects of a nuclear accident, was on the spot. Unable to make a decision, the technical leaders of the NRC turned to Barrett to answer their fears. It looked as if it was time to move people out of the area around Three Mile Island, but how far away should they be moved? Barrett would know, Barrett had been right so far, he must know the answer!

"I'm not sure," Barrett said. "I can't tell you for sure, but ten miles is more than enough, ten miles is plenty . . ."

Ten miles was also twice the distance Civil Defense directors had been making plans for since Wednesday. These thrown-together plans all were designed around the idea of taking residents who lived within five miles of Three Mile Island and moving them the short distance into the area from five to ten miles from the plant—just the area that Lake Barrett had suggested should also be evacuated. This greater radius included nearly four times the area, five times as many people, and for the first time, hospitals and nursing homes. If the NRC recommended an evacuation to ten miles, Colonel Henderson's work would be cut out for him.

One person in the meeting room *was* familiar with the Pennsylvania emergency plan. Harold Collins was with NRC's office of state programs. He knew Oran Henderson and should have known that Pennsylvania was only ready to evacuate to five miles. If that. But Harold Collins—called "Doc" Collins—was as dazed as the rest. In one motion they had completely by-passed discussion of *whether* to order an evacuation, in favor of deciding just how far people should be moved.

"It was my decision to make if it was really coming out," Denton said, "and if it was that high and had a rem an hour, it's an awful lot of radiation coming from a plant. It was important to evacuate quickly, because as the plume would begin to drift over populated areas, you couldn't wait two hours or five hours for a careful analysis. The exposure would already be there . . . I was making decisions in the face of uncertainty."

The decision was made. There would be an evacuation. No checking with the NRC commissioners, with Governor Thornburgh or the

White House. Doc Collins called PEMA with the evacuation recommendation.

The NRC's own procedures would have required Collins to contact Tom Gerusky at the Bureau of Radiological Protection, but Collins decided instead to by-pass a middle-man like Gerusky, who would probably have vetoed an evacuation, in favor of contacting Oran Henderson, who was ready to begin an evacuation at a moment's notice. Collins mentioned this change of procedure to Denton and the others. As in every other decision made in the room that morning, there was no dissent.

In their telephone conversation, Henderson told Collins that the radiation had been measured by a helicopter 600 feet above the plant, and not off-site as Denton and the others had guessed, but that did little to change the message from Bethesda.

COLLINS: You've notified the locals to be prepared for an evacuation?
HENDERSON: Yes.
COLLINS: . . . we recommend that you evacuate people out to ten miles from that plume, in the direction of the plume.
HENDERSON: Well, we'll start with five maybe.
COLLINS: That is, of course, your option, but I would certainly start with at least that, and you'd better start thinking of moving from five to ten.

This was the evacuation that Oran Henderson had been expecting all morning. He called Lieutenant Governor Scranton who, in turn, called Governor Thornburgh. Thornburgh would have to pass judgment on any evacuation recommendation.

Henderson called the Civil Defense directors of Dauphin, Lancaster and York counties to say "We have about a ninety-percent chance of conducting this evacuation, so I want to make certain that you are in as good a posture of readiness as you can be."

The three counties, already on alert, began to activate police and fire units in preparation for any move. In Dauphin County, Kevin Molloy went even further. He warned the public.

When Henderson talked to Kevin Molloy, the PEMA director left

161

the impression that an evacuation was about to take place and that Molloy would be receiving a confirmation call from the governor within a few minutes. Rather than waste time, while waiting for the call from Thornburgh that he was sure to get, Molloy had his dispatchers put all local fire companies on alert; they called local schools, ordering children indoors. Windows were closed and school buses summoned to evacuate the children. Then Molloy took the process of preparation a step further by warning the people of Dauphin County over one of the local radio stations.

"I called WHP," Molloy said, "and I went on the air and just stated that, as a result of the incident at Three Mile Island, the possibility does occur that we might have to take some type of protective evacuation. If you have to leave, we'll let you know over the radio . . . but don't do anything until we give you the word."

Jim Moyer was doing the news at WHP when Molloy called. "Well, I answered the phone and there was no question that he was going to get on the air," Moyer remembered. "He made his announcement, and then I did a relaxed sort of question-and-answer with him and by then my heart was pounding pretty hard. After he finished, we continued to run the story using tape from the announcement and the interview . . ."

Having made his announcement to the people of Dauphin County so that they might be ready for the impending evacuation, Molloy waited for the call from the governor that Henderson had told him to expect any minute. It never came.

If WKBO's coverage of Dr. Sternglass on Thursday had caused panic in Harrisburg, Kevin Molloy's announcement on WHP caused hysteria on Friday. Within the hour after Molloy made his announcement, more than 137,000 telephone calls were made in Harrisburg—ten times the normal load and more than two calls for each man, woman and child. The system was jammed and for most of the late morning it took 10 to 15 minutes just to get a dial tone. Many people simply got in their cars and left the area. Lines appeared at service stations and at banks. One woman in York, twelve miles from the plant, withdrew $100,000 in savings bonds from her safe-deposit box, afraid they would be contaminated. The Federal Reserve sent an additional $7 million in cash to local banks to cover panic withdrawals.

An automobile dealer in Harrisburg sold twenty-one cars on Friday—the highest one-day sale in his dealership's history. "Nobody even wanted to haggle," he said, "and a couple of them even paid with cash money. You know, tens and twenties." Most of the twenty-one cars were bought by doctors—some of them physicians who had signed agreements with Met Ed to treat victims of radiation sickness in the event of a major accident at TMI.

There was some panic in the schools. Barb Taylor went to help another teacher at Herbert Hoover Elementary School in Harrisburg. When she returned to her fourth-grade classroom, she found the children writing their wills. "My bed go to my sister Susie and my soul go to heaven," wrote one little girl.

While Harrisburg and Middletown were in turmoil, the technicians in the Unit 2 control room worked on calmly, not knowing anything of the panic in their community, among their families and friends. Venting the waste gases had dropped pressure in the makeup tank, just as Gary Miller had hoped. But Miller knew the cause of the pressure had not gone away and the whole procedure would have to be performed again . . . and again. But at least this first time had caused only a short release of radiation and the valve had not stuck open.

This comparative serenity was destroyed around 10 A.M., when an auxiliary operator walked into the 40-by-40-foot control room to announce "I got a call from my wife. She heard on the radio that the NRC is ordering evacuation downwind. She's going to school to get our kids out of here."

Chick Gallina, who until this point had thought that Bethesda would not make any recommendations without checking first with him, couldn't believe what he was hearing. "You've got to be kidding," he said. No one was kidding.

Gallina turned to Floyd and Miller. Was there something wrong at the plant he didn't know about? Was there a big radiation release coming up? No, things are fine.

A call came in for Gallina from Tom Gerusky of BRP. The state wanted to know where Doc Collins had gotten the idea that an evacuation was in order.

"They didn't get it from us," Gallina responded. "I don't know where they got it. We don't think there is a need for evacuation."

"I was mad," Gallina said later. "They were circumventing the licensee's procedures and circumventing the state's procedures by doing that. I was mad at the NRC headquarters. I was really livid. The procedures say that . . . they notify the Bureau of Radiological Protection, so you have two intelligent people who at least know the jargon and meaning of the stuff, and if they agree, then the site or the state may make a recommendation and they go to the governor for the order . . . but it was done directly from the NRC headquarters, which has no business recommending evacuation of anyone, and it had gone to Civil Defense, which doesn't know what these numbers mean, and it caused a panic. I said, 'We've got to stop it!' "

Gallina followed normal communication procedures and called George Smith in King of Prussia, the same George Smith who had learned to keep his mouth shut the night before when he pointed out the error NRC was making in ordering Miller to stop dumping wastewater. "I was thoroughly disgusted," Gallina recalled.

"George, we can't let this happen!" Gallina yelled into the phone. Around him, a gaggle of utility engineers were yelling at the NRC investigator: "What are you guys doing? Do you realize what it is going to do to us and to the industry? Do you realize what it is going to do to the people around here?"

Smith thought the evacuation order had come from Hendrie and the other NRC commissioners. "Chick," he said, "it's a management decision. I don't agree with it. I agree with you a hundred percent. This is uncalled for, but . . . we can't say anything about it."

As Kevin Molloy was waiting for his official evacuation order, Colonel Oran K. Henderson was doing his best to get it for him. Henderson had been expecting to evacuate *even before* the call from Doc

Collins. Collins' call just clinched it in the PEMA director's mind that people would be moving that Friday. But the governor must approve any evacuation order, so Henderson called Scranton, who would call Thornburgh, while Craig Williamson, Henderson's deputy, called Gerusky. Word of Collins' call flashed through the state government.

Bill Dornsife and Maggie Reilly couldn't understand what was going on. They saw no need for an evacuation. Outraged, they telephoned Collins to find out who was responsible for the call. Maggie Reilly was a strong-willed woman of Irish descent who had worked her way up through the state government for fifteen years. She was a fighter and she was furious. Doc Collins and the NRC were no match for Maggie Reilly.

REILLY: Name some names!

COLLINS: I don't know whether I should, Margaret. You don't really need to know names at this point in time . . . We did what we were told. Okay. Then I called. Then I called you. All right, then Henderson, you know, said—Then we got in touch with him later and he said that the—you folks didn't agree with him. And he wanted to do something. He wanted to evacuate, so, you know, I guess maybe he's making up his own mind too as to what he wants to do. So you know, you got communications and organizational problems I guess on both sides of the fence.

REILLY: Welcome to the club.

COLLINS: Yeah, right. So—

REILLY: Sorry about getting nasty, Doc, but that was a low blow for those turkeys.

COLLINS: These people are way uptight about this whole thing. They don't want to—They don't want to—They don't want anything to go wrong. They're—They're using terms like precautionary evacuation.

REILLY: Right.

COLLINS: And they'd rather be safe than sorry, you know. They— They would much rather be safe than sorry, and I think this is what motivates them, you know. So there's no hard feelings between you and me, I hope.

REILLY: No, I'll get over it.

165

COLLINS: Yeah, okay. So I didn't do this on my own, I want you to understand this.

The state wasn't at all sure that Collins hadn't made the call on his own. He'd called the wrong agency by calling PEMA and then refused to tell Reilly and Dornsife who had ordered him to make the call. About the only piece of useful information the two had been able to glean from Collins was that none of the NRC commissioners knew the recommendation to evacuate had been made.

Collins was shaken by his encounter with Maggie Reilly, but still resolved. An evacuation *should* take place. To ensure this, he placed a second call to Henderson at PEMA. "Ken," Collins said, "I just want you to know that that was not only my recommendation, but that it has the backing and support of the commission."

Governor Richard Thornburgh was faced with an important decision: Should he take the advice of unnamed experts at the NRC, conveyed to the wrong agency within the state government by a man he had never heard of?

If there was a real threat to public health posed by the 1,200-millirem release, then Harold Denton knew that moving people out of the area was the first priority. Time was important—so important that rash decisions were being made and, later, defended. "The important thing is to get out there and get people moving before the plume gets there," Denton said. "We didn't have time to do an elaborate evaluation and so forth."

Richard Thornburgh didn't have time to do an elaborate evaluation either, yet he ordered one. Given the unknown source of the evacuation recommendation, Thornburgh's growing distrust of the NRC and his adversary method of decision-making, the governor was unable to make a quick decision.

The governor began making calls himself, first to Scranton, then Henderson. "I told him exactly what Henderson had told me," Scranton claimed. Thornburgh called Henderson to ask about Collins. "I told him that we had worked together before," Henderson said later, "that Collins enjoyed a good reputation, that I did not personally know him except through discussions in the office."

Thornburgh, who two days before had not known Henderson at all, asked the retired infantry colonel for his recommendation. "Governor," Henderson said, "since I've not heard yet from BRP, I have no alternative but to recommend that we conduct a five-mile evacuation."

Thornburgh hadn't heard from BRP either. Gerusky, Reilly and Dornsife were all trying to call the governor's office, but there were no free telephone lines in Harrisburg. Long-distance calls were no problem, but calling across town was nearly impossible. Eventually, Gerusky started out on foot for the state capitol to have his say.

A meeting was already under way in Thornburgh's corner office in an attempt to come to some decision. Thornburgh, Waldman, Critchlow, Scranton, MacLeod and Knouse were present in the gold-carpeted office, lined, like all the rooms in the executive suite, with portraits of former governors above the richly carved oak paneling. Facing Thornburgh at his desk was a marble fireplace and above it a portrait of William Penn. Windows at the governor's back faced in the direction of Three Mile Island.

There was little information for the group to work with. They knew that there had been an uncontrolled release of radiation from TMI; that there had been a 1,200-millirem-per-hour reading made 600 feet above the stack; that a Met Ed shift supervisor had talked of evacuating noncritical personnel from TMI, but that this concern was apparently not shared by other personnel in the control room; that both DOE and BRP had confirmed a release but that both agencies were skeptical of the necessity of evacuation; and, finally, that a little-known NRC official named Collins had called the wrong state office, recommending a five- or ten-mile evacuation downwind, and refused to indicate who among his superiors had authorized the call. They also knew that Henderson had recommended a precautionary evacuation and the people of Dauphin County were already alerted. Health Secretary Gordon MacLeod was concerned about unborn children and favored some action to protect pregnant women.

At Thornburgh's request, Paul Critchlow asked Karl Abraham to check out Collins and the evacuation recommendation with Bethesda. Abraham's call to Bethesda wasn't very much help to Thornburgh and Critchlow. He did determine that Collins had made the call to Henderson, recommending evacuation, but when asked whether Collins was authorized to make such a call, Abraham could only say, "I don't think so."

Then, fifty minutes after the NRC had first recommended evacuation to ten miles, with panicked citizens leaving town in reaction to Kevin Molloy's announcement, which was still being played on WHP, Harrisburg's air-raid sirens went off without warning.

"What the hell is going on!" shouted Thornburgh.

No one has ever taken credit for the five-minute warbling blast that sounded just like all those cold war tests, but it did clear the streets. And on the floor of the state assembly, the sirens produced a rare moment of complete silence.

Tom Gerusky pounded up the hundred limestone steps to capitol hill, out of breath from his solo sprint across town, just as Thornburgh decided to call Joseph Hendrie, chairman of the NRC, and find out what he knew.

Jack Herbein and Walter Creitz met the press in Middletown that morning. After their problems on Thursday at the Hersey Motor Lodge, the utility executives tried a change of scene to the American Legion Hall, three miles up the road from TMI. Two hundred reporters and cameramen were there, most of them angry at not having been told of the wastewater release the day before. Herbein still wore the dark suit and striped tie he'd worn Wednesday and Thursday.

The reporters laid into Herbein for not telling them about Thursday's wastewater release. Shouting their questions, some of them standing on the folding chairs rented from a local funeral parlor, the reporters confronted the two men who stood on a small stage, while the rest of central Pennsylvania was listening in. WHP radio had run a microphone wire across the alley next to the Legion Hall and into the home of a friend of one of the station's engineers. That friend's phone carried the press conference to five radio stations in the area.

"It was insane, the yelling and screaming," remembered Joel Shurkin of the Philadelphia *Inquirer*. "Stuck at the back of the hall, I could hardly even see, much less hear, what was happening. So I turned on my transistor radio and heard what was going on that way."

When finally asked about that morning's radiation release, Herbein said it was more like 350 millirems than 1,200 and there was nothing to worry about. Mayor Bob Reid, wearing a Middletown High football jacket, turned to his friend John Garver and asked, "Ain't this the shits?" Neither man believed what Herbein was saying. In fact, nobody in the room believed it.

"What about the workers?" shouted Jimmy Breslin of the New York *Daily News*. Breslin was obsessed with the workers who had received overdoses of radiation on Wednesday. "Bring out the workers!" he yelled at Creitz. "I bet ya can't bring 'em out 'cause their balls fell off!"

The press was demanding information about the disaster that they were sure had come and Herbein was insisting that there was no problem. "But why didn't you tell us?" screamed a reporter. "I don't know why we need to tell you each and every thing we do," replied Herbein. The press does not like to hear that kind of talk.

"I hate gang bangs," said Curtis Wilkie of the Boston *Globe*. "This was much worse than the campaign trail, or a Democratic convention or a New York press conference, which is the worst thing I can say about anything. You have twenty-seven action news teams with all their cameras, needing to be on the air with their questions and knocking each other over the head. TMI was hideous. Most of them appeared to be just as ill-informed about the subject as I was."

And while Creitz and Herbein were assuring the press that there was no problem, no threat to public health, no need to evacuate anyone, the Middletown police cruised outside the Legion Hall, using bull horns to tell people to clear the streets, go to their homes and shut all windows.

When the commissioners of the NRC met together in Washington that morning, the five men felt generally confident that the accident was ending. But Harold Denton, who came in from Bethesda for the briefing, harshly told them that the NRC was not on top of the situation. Denton's greatest concern seemed to be the poor communication between Bethesda and the site. "We do have our people in the control room who search out the answers," Denton said. "But with regard to an actual or hard number for release, rate, curies, quantities off-site, that process seems to take hours."

Both NRC Chairman Hendrie and Governor Thornburgh were suffering the effects of the communication problem Denton mentioned. As the final authorities in their respective areas, the two men were feeling pressured to make decisions without adequate information. And each wanted now to talk with the other. "His [Thornburgh's] information is ambiguous, mine is nonexistent," Hendrie said. "I don't know, it's like a couple of blind men staggering around making decisions."

Denton was still in favor of a "precautionary evacuation in front and under [the plume]. But others thought it would be enough to recommend that people remain indoors and Hendrie agreed, expressing concern that people might evacuate *into* the path of the gas plume.

When Gerusky finally ran panting into the governor's office, he found Thornburgh and his advisers surrounding a speakerphone, debating with Hendrie about what action to take. The plume of radioactive gas that had measured 1,200 millrems per hour in the helicopter nearly an hour before was long gone. The release was practically forgotten in the Unit 2 control room, but it still seemed very real in Harrisburg and in Washington, where the NRC commissioners gathered around their own speakerphone to consult with the governor of Pennsylvania.

". . . It appears to us that it would be desirable to suggest that people out in that northeast quadrant within five miles of the plant stay indoors for the next half-hour," Hendrie told the governor.

"So your immediate recommendation would be for people to stay indoors?" asked a surprised Thornburgh. ". . . Was your person, Mr. Collins, in your operations center, justified in ordering an evacuation at 9:15 A.M. or recommending that we evacuate at 9:15 A.M. or was that based on misinformation? We really need to know that."

"I can't tell what the—I can go back and take a check, Governor, but I can't tell you at the moment," Hendrie replied. "I don't know."

Hendrie hedged in answering the governor's question about Collins. Thornburgh's main concern was with Collins' *authority* to call the state on behalf of NRC. Hendrie knew that Collins had been given authority to do so by Denton, but he chose to avoid this question in favor of the more philosophical, and less incriminating, question of whether the information available justified the recommendation. That was enough for Thornburgh. It seemed to him that Hendrie didn't even know who this fellow was, so how could Collins claim to speak for the NRC?

At the end of Thornburgh's five-minute conversation with Hendrie, Tom Gerusky finally got to say what he had run across town to tell the governor. By that time, Gerusky had been joined by Dornsife, Craig Williamson of PEMA and Randy Welch of the State Health Department, all eager for the governor not to order an evacuation.

The decision was made. There would be no evacuation on Friday. Still, some action had to be taken to make it look like things were running smoothly in Harrisburg and Washington, to deal with what Craig Williamson of PEMA called "the position that the governor was in."

A few minutes after his conversation with Thornburgh, Joseph Hendrie received another important call. Jimmy Carter wanted to know what this latest ruckus was at Three Mile Island. NRC Commissioner Victor Gilinsky, the most politically oriented of the five commissioners, had been making regular calls to Jessica Tuchman Matthews of the National Security Council. Gilinsky's call that morning had been an alarming report of a "serious uncontrolled release." Matthews told Brzezinski, who went straight to the President. And now Carter turned to Hendrie for some sign that the NRC was on top of the situation.

Hendrie explained the current situation as best he could and complained about "savage communication problems." Carter promised to help. He immediately ordered the White House Communication Agency to have special phone lines installed between the White House, NRC and Harrisburg; he asked Brzezinski to look into the advisability of the NRC's taking over direct operation of the nuclear plant—an action already rejected by the NRC—and he called for a helicopter to take Denton to Pennsylvania.

Carter then tried to reach Governor Thornburgh and found himself stricken with the same problem as everyone else trying to call Harrisburg that morning—the number was busy.

Thornburgh was still feeling pressure to do something to legitimize the morning's overreaction, both in Bethesda and Harrisburg. Speaking a second time to Hendrie, he advanced Health Secretary MacLeod's idea of the previous day that pregnant women and small children should leave the area—a recommendation he had publicly dismissed the day before when Dr. Sternglass first suggested it.

"If my wife were pregnant and I had small children in the area, I would get them out because we don't know what is going to happen," Hendrie responded. "I go along with you on that, Governor, and I think there ought to be an evacuation."

"What are you talking about distance-wise?" Thornburgh asked.

"Two or three miles," the NRC chairman said.

"At that point, it got down that two or three miles was ridiculous," Gerusky said. Still an idealist, Gerusky did not support the idea of moving women and children and upsetting the community just so the state would not look bad. As Thornburgh and Hendrie decided on the details of the evacuation, Gerusky sat in the governor's office, his head in his hands, slowly rocking.

Craig Williamson remembered, "I thought momentarily, 'My God, Tom, you'd better get yourself organized and support the governor.' And then he raised his head out of his hands and indicated to the group that he understood."

Gerusky later said, "I came away feeling that it was a cover-your-ass recommendation."

Five miles was accepted as the logical distance for any sort of evacuation. Then there was a problem of what age limit should be suggested for evacuating children. "Infants" was unclear. Was a three-year-old in less danger than a two-year-old? It was finally decided that "preschool" would cover the young children well enough.

It was decided that the governor would "order" the schools to close, but only "advise" that pregnant women and preschool children evacuate. That way, should a departing mother or child be injured, the state's liability would be limited.

A statement was issued from the governor's office at 12:30 P.M.:

"Based on the advice of the chairman of the NRC, and in the interests of taking every precaution, I am advising those who may be particularly susceptible to the effects of radiation—that is, pregnant women and preschool children—to leave the area within a five-mile radius of the Three Mile Island facility until further notice. We have also ordered the closing of any schools within this area. I repeat that this and other contingency measures are based on my belief that an excess of caution is best. Current readings are no higher than they were yesterday. However, the continued presence of radioactivity in the area and the possibility of further emissions lead me to exercise the utmost of caution."

Brian Grimes of the NRC looked ahead to further problems with the evacuation recommendation. "It concerned me very much at the time that we would move people for no reason and then not have a criterion to bring them back in. We might move a large number of people without a significant reason. This situation could go on for many days. What would be the criterion for telling them it's safe to come back in?"

Tom Gerusky walked down the capitol steps on his way back to the Department of Environmental Resources. He'd just given his agency's approval to a move that he knew was unjustified and completely political. He walked through the wooded capitol grounds and out to the street. It was a beautiful day in early spring and Gerusky noticed there was no one else on the street, no cars were moving. There was no sound in the normally bustling city of 60,000. "I wondered where everyone was," he said. They were inside, with doors and windows closed, as ordered. And they were scared.

17

Possibility of a Meltdown

Over Friday afternoon and evening, more than 40,000 people left the area around Three Mile Island. There was no order for them to evacuate. They left because they were afraid, and they went in all directions. One of the routes followed Highway 39 through Hummelstown, where a sign in front of a shop said, "Come in out of the rays!"

The airline flights that normally turned low over TMI on their final approach to Harrisburg International Airport were all canceled. So was the Met Ed employee Easter egg hunt that had been scheduled for Saturday on the observation center lawn. People were already beginning to call the day Black Friday.

Four-year-old Amy Sanders admonished visitors to her mother's dress shop, "When you go out, don't breathe." Later, when walking down the street, she apologized: "I opened my mouth wide, Mommy."

Amy's mother, Jessie Sanders said, "When they built it here, I was ignorant about the dangers. We never realized how dangerous it was. Person after person now says to me, 'If there's an antinuclear club, I'm a member.' We're volunteering for an organization we don't even know the name of."

The quiet, steady, hard-working people of Middletown and Goldsboro were quickly sending their children out of the area. "We were besieged by parents coming up here to take their children home," said Wendell Poppy, principal of the Londonderry Township Elementary School. "Many families told us they were going to stay with relatives or leave the area."

"If the effects are going to show up twenty years from now," said Mrs. Sherry Frenndel, who lived near Middletown, "I don't think it is worth keeping my little boy around. I'm disgusted. They waited hours before telling us about the accident. They haven't told us when it will stop."

"I'm afraid of what might happen next," said Mrs. Margie Borelli of Middletown. "We don't know how serious this may be. We don't know if the plant will blow up."

Tolltaker Robert Haines watched a steady stream of cars leave Harrisburg, passing through his gate to the Pennsylvania Turnpike. Soon Bob Haines decided to leave, too.

"Shall I sell my stock in Metropolitan Edison?" asked a bartender in Harrisburg.

Nine-year-old Kim Hardy was bused from school back to his home in Etters within sight of the cooling towers: "Some of the kids started to cry," he said. "I didn't cry, but it was sure hot on that bus with all those windows up."

By 5 P.M. Annette Baker of Goldsboro noted, "Almost everybody's gone. Normally at this time of day, the people are around, coming home. I don't mean this place is a traffic jam or anything like that. We've only got six hundred or so people. But there's just nobody here."

"I guess I'll have to leave, too, if they say I should," remarked farmer Joseph Conley of Yocumtown. "But I don't know what I'll do about my cattle or this place. My father built this house and I was born here. I don't think I could even sell it at this time. I don't know what it's really worth. But I sat down and figured out it would cost $250,000 to replace it . . ."

"They're not saying much to us," said a teller at the Dauphin Deposit Bank. "They just take their money and leave."

By evening Mayor Kenneth Myers of Goldsboro said, "I feel like I'm the mayor of a ghost town."

*   *   *

President Carter had asked Hendrie to send a high-level official of the NRC to Pennsylvania. Since Harold Denton was already going, the chairman decided that he would do just fine. And when Governor Thornburgh asked Carter for a presidential representative, the President replied, "I am sending you Harold Denton of the NRC as my personal representative for the duration of this problem."

As Denton drove his 1978 Dodge Omni to Bethesda Naval Hospital, where the helicopter Carter had sent waited, he had no idea how much was expected of him by Thornburgh and the White House. Next to the hospital helipad was the school where Mrs. Denton taught. She heard the beating sound of the helicopter rotor as her husband left for the crippled reactor. When Denton stepped out of his blue presidential helicopter in a field next to the observation center that afternoon, he thought his job was just to make things run a little smoother than they had. Instead, Denton found he was expected to be a messiah, bringing efficiency and truth where there had been little. The balding, overweight engineer from North Carolina was proclaimed "doctor," though he held no Ph.D. And if anyone in Pennsylvania had been told that Denton had recommended evacuation that morning, they would not have believed it.

There were two stories to cover in central Pennsylvania that afternoon. There was the obvious story of the public reaction to Molloy's announcement on WHP, the governor's evacuation recommendation and the panic coming from the NRC in Bethesda and from the state capitol. That was the easy story and the route most reporters took. Then there was the story of the Collins call, the *reasons* for what had happened in Bethesda and Harrisburg—a story that would not be fully understood for months.

John Fialka of the Washington *Star* heard from congressional sources that the NRC had recommended evacuation Friday morning. Fialka's story was the only one to place responsibility for the events of the morning with the NRC, but it was lost in the crush of events on Friday night.

Until Denton's arrival, an important sidebar had been the unreliable and contradictory information put out by the utility, the NRC and the state. No sources were trusted. But then Denton stepped out of his helicopter, mumbled half a dozen words to reporters and was immediately accepted by both the press and Thornburgh as an honest and reliable source.

Steve Liddick of WCMB radio tried to explain the Denton phenomenon: "Harold Denton was trusted because he looked like a regular, down-to-earth kind of guy. And people wanted someone to believe. It was just like after Richard Nixon resigned—the public was ready to support anyone at all. Attila the Hun could have come in here and done really well."

An army of reporters decided to emphasize the colorful aspects of the story, sometimes seeing more color than was really there. One television crew asked Middletown citizens to stay out of the street scene they were shooting. That night the footage was shown as "the deserted streets of Middletown, Pennsylvania." Another crew was said to have moved a FOR SALE sign from house to house, to show panic selling of real estate.

In fact, there was some panic selling—or at least people offering their homes for sale. And the streets often were near deserted.

Curtis Wilkie of the Boston *Globe* pointed out to his readers that when the Pennsylvania Dutch put hex signs on their barns to guard against evil, they didn't know they'd need one against radiation. A Japanese reporter from Hiroshima asked local residents if they knew what had happened to his city during World War II. Most people under thirty did not know.

In the afternoon, the state and Red Cross set up a shelter for evacuating women and children at the sports arena in Hershey. Mothers, children and whole families arrived to spend the night on cots set on the gym floor. Cardboard boxes were used to build walls to create privacy—privacy that the basketball court and the media would never allow. Most families didn't go to the arena. They went to relatives and friends. All were waiting for word from the governor that danger had passed, that it was safe to return to their homes. By early Friday evening, however, Thornburgh's attention was diverted, for there was a new problem at Three Mile Island.

*　　　*　　　*

177

Two days before, on Wednesday afternoon, Ed Fredrick had watched the needle on his containment pressure gauge jump from 2 to 28 pounds per square inch. He thought the momentary fluctuation —a "pressure spike"—was a problem with the instrument. Maybe a voltage surge. Nothing important. The incident was forgotten as the inked tracing rolled around the instrument's recording drum, out of sight and into the history of the reactor. With the core uncovered and no cooling pumps in operation, there were more important concerns in the Unit 2 control room.

But sometime on Thursday night or early Friday morning, that 28-pound pressure spike became extremely important in Bethesda.

Under scrutiny by experts at the NRC, what had been only a bump on the paper tape—heard as a "thud" or a "slamming valve"— proved to be a hydrogen explosion. With the verdict in from the primary-coolant sample, and this pressure spike, it looked like the zirconium cladding had reacted with water to release hydrogen gas in the reactor. Some of the gas had escaped through the pressurizer and exploded with air in the containment, creating the 28-pound pressure spike. And the rest of the hydrogen was still in the reactor, gathered in a bubble at the top of the pressure vessel. A very dangerous bubble.

Only ten minutes after Thornburgh announced his recommendation that pregnant women and preschool children leave the area, the NRC commissioners heard by telephone from Roger Mattson in Bethesda that the reactor core had been uncovered for some time, resulting in "failure modes the likes of which have never been analyzed." Mattson, director of the NRC's Division of Systems Safety, had been pressed into service analyzing the condition of the reactor core on Thursday afternoon. As if news that the core had been uncovered were not bad enough, Mattson also had calculated that there was a bubble of hydrogen gas filling more than a thousand cubic feet in the top of the reactor vessel. If system pressure dropped, the bubble would expand and uncover the core again.

"They can't get rid of the bubble," Mattson told Hendrie. "They have tried cycling and pressurizing and de-pressurizing. They tried natural convection a couple of days ago, they have been on forced circulation, they have steamed-out the pressurizer, they have liquided-out the pressurizer. The bubble stays."

Mattson had decided there was no way to bring the reactor to a cold shutdown with the hydrogen bubble in place and he saw no apparent way to get rid of the bubble.

"We have got every systems engineer we can find, except the ones we put in the helicopter, thinking the problem . . . and they are not coming up with answers. We have got the Navy working . . . B&W is in constant communication with GPU decision-makers at this point. We don't have a solution, but maybe they are coming up with one."

"It sounds to me like we ought to stay where we are," Hendrie said. "I don't like the sound of de-pressurizing and letting that bubble creep down into the core."

"Not yet," Mattson agreed. "Not yet. I don't think we want to de-pressurize yet . . ."

But Mattson was worried about the possibility of a meltdown if the bubble were to expand. "I'm not sure why you are not moving people," he said. "Got to say it. I've been saying it down here. I don't think we are protected at this point. I think we ought to be moving people . . . It's too little information too late, unfortunately, and it's the same way every partial core meltdown has gone. People haven't believed the instrumentation as they went along. It took us until midnight last night to convince anybody that those goddamn temperature measurements meant something. By four o'clock this morning, B&W agreed."

Just forty-eight hours before—a time that seemed on Friday to be weeks or months before to the tired experts of the NRC—a technician's hand-calculated core temperatures had said the fuel was nearly melting. And no one had believed it.

Intentional de-pressurization of the cooling system by the Met Ed engineers was very unlikely, but that said nothing of an *accidental* de-pressurization, which seemed increasingly likely to take place. The tenuous stability of the primary cooling system relied on the only one of four main coolant pumps to remain in service. There was no backup. And that last pump was beginning to show definite signs of strain.

"We've never had a pump working in such high radiation," said Edson Case. "It's way beyond what it was designed for, and we just can't be sure how long it will last. The tech specs allow the main pump shaft to wobble no more than ten thousandths of an inch, and we're already wobbling more than thirty thousandths and getting worse."

Commissioner Victor Gilinsky asked Mattson what would happen if the pump failed.

"Hours," Mattson said.

"Hours before what?" Gilinsky asked.

"Hours before you had a core melt," said Mattson.

The actual calculations said 200 minutes—three hours and 20 minutes—before fission products might be released and what newswoman Cristine Russell had called "invisible radioactive fragments" could enter the lungs of those who lingered around Three Mile Island.

At 1:30 P.M., Joseph Hendrie stood at a huge conference table in the Situation Room buried beneath the White House. Summoned from NRC headquarters on H Street, he faced representatives of the White House, Department of Defense, Joint Chiefs of Staff, National Security Council, Department of Energy, Department of Health, Education and Welfare, and a dozen other federal agencies. It was an imposing group to be confronted by the chairman of what one reporter had called a "dinky little asshole agency." Backed by detailed maps of the world and a view of the men and machines who monitored that world for the nation, Hendrie told the group his fear for Three Mile Island. He said that it might melt down.

After the slender white-haired Hendrie sat back down, Zbigniew Brzezinski suggested that Jack Watson be placed in charge of coordinating the federal response to TMI. There was no dissent, and Watson took over the floor. A member of the "Georgia Mafia" that came with Carter into the White House, Watson was a lawyer and

180

one of Carter's top domestic trouble-shooters. His orders were decisive, to the point, probably necessary at that moment—and very heavy-handed:

1) Harold Denton would be the sole source of information for federal agencies dealing with the crisis and for the press; 2) All press releases and contacts dealing with federal involvement would be handled by Jody Powell in the White House, with the exception of certain technical questions that might still be answered by the NRC; 3) Paul Critchlow and Governor Thornburgh would be encouraged to deal only with questions of public safety and emergency preparedness; 4) Robert Adamcik, regional director of the Federal Disaster Assistance Administration, would be sent to Pennsylvania to coordinate evacuation preparations.

The time had passed for pleasantries. Now the White House would intercede where it thought best, whether with the NRC, Commonwealth of Pennsylvania or Metropolitan Edison.

While Hendrie was at the White House, more calls were coming into the NRC commission room on H Street, but the other four commissioners present seemed unable to deal with the situation in Hendrie's absence. They could not make a decision without the only commissioner with a technical background.

Mattson had called again from Bethesda. ". . . my principal concern," he told Commissioner Gilinsky, who was second in rank to Hendrie, "is that we have got an accident that we have never been designed to accommodate, and it's, in the best estimate, deteriorating slowly, and on the most pessimistic estimate is turning bad. And I don't have a reason for not moving people. I don't know what you are protecting by not moving people."

Gilinsky shared Mattson's concern and wanted Doc Collins to call Henderson to make sure the state was ready to mount a full-scale evacuation around the plant, but Commissioner Kennedy resisted. "It is going to be in the newspaper this evening at five o'clock: 'NRC contemplating evacuating.' If that's what you want, all right," Kennedy said.

Collins was told to call anyway. Reporting back to the commis-

181

sioners a few minutes later, he was asked by Gilinsky how much time would be required to evacuate the ten-mile area. Collins had forgotten to ask, but made a guess:

"My thought on it," Collins said, "just sort of a gut feeling . . . I would think they would be able to get them out inside of an hour, at the most two."

"Are you talking about Harrisburg, too?" Gilinsky asked.

"I would have to include that," Collins answered, "if it went toward Harrisburg . . . I would say they ought to be able to get all the small towns in the counties and the local folks out within an hour, and probably certainly have the city cleared by two or, you know, something on that order. I'm estimating, since I don't live there, I really don't know . . . It's a difficult question, Commissioner."

Collins not only had no idea how long an evacuation would take, even his "gut feeling" was far off. The Harrisburg police, who had managed evacuations during floods, had determined it would take at least twenty hours to do the job and said so in their flood-evacuation plans written long before Three Mile Island. Collins hadn't thought of the three hospitals, four nursing homes, two jails and a prison that would be included in an evacuation of Harrisburg. His oversight left Gilinsky and the other commissioners feeling that there was time to get people out of the area if the situation seriously deteriorated.

When Thornburgh first heard about the presence of a hydrogen bubble at 3 P.M., he asked Hendrie if it could explode, rupturing the pressure vessel and releasing radiation. But Hendrie told him that there was no oxygen in the pressure vessel to burn with the hydrogen gas and no source of a spark to set it off. While the governor found this news reassuring, Hendrie also suggested that emergency workers be placed on alert and that "if we suspected getting a fairly husky release, the evacuation might extend twenty miles."

Once again, Oran Henderson's evacuation plan was obsolete. Now he would have to start over, looking for new routes, new shelters and more buses to move the ever-larger group of potential evacuees within the twenty-mile zone.

"Is there anyone in the country who has experience with the

182

health consequences of such a release?" Thornburgh asked.

Hendrie admitted there was no one available.

The governor then talked to Denton, who was at the plant. Denton seemed less upset than Hendrie had been and Thornburgh was somewhat reassured. The two men agreed to meet at 7 P.M. to discuss conditions in the reactor.

While Thornburgh was talking to Denton, Jay Waldman, the governor's chief aide, heard from Jack Watson at the White House. "Jack Watson asked me if we could please not request the President to declare a state of emergency or disaster," Waldman recalled. "He said that it was their belief that that could generate unnecessary panic, that the mere statement that the President has declared this area a disaster area would trigger a substantial panic; and he assured me that we were getting every type and level of federal assistance that we would get if there had been a declaration. I told him that I would have to have his word on that, an absolute assurance, and that if it were true, I would go to the governor with his request that we not formally ask for a declaration."

Beyond the issue of of panic, Watson was very concerned that declaring a state of emergency would spell an end to the commercial nuclear-power program that his President supported and hoped would play a substantial role in making the nation less dependent on imported fuels. And, despite Watson's assurances to the contrary, the Air Force had already begun to withhold promised air transportation in fear that compensation would not be forthcoming without the declaration.

Watson would later deny that he asked Waldman to intercede with Thornburgh in this matter. One of the men lied.

Commissioner Victor Gilinsky of the NRC was worried about the image of his agency and the lack of public information coming from Bethesda. If the NRC was doing good work, then they should be getting good press, it seemed to him, but hardly any stories were coming from Bethesda, where a number of reporters had been trying to get information with only limited success. Robert Schakne of CBS and Walt Mossberg of *The Wall Street Journal* were continual call-

183

ers. At one point on Friday night, an NRC engineer complained, "I've just hung up on Mossberg for the third straight time, and now he says it's a matter of critical national importance!"

So while Jack Watson and Jody Powell were doing their best to limit the number of different information sources available to the press, Victor Gilinsky asked that another source be opened. He went to Frank Ingram, the NRC's number-two public information officer, who was in charge in Bethesda after his boss, Joe Fouchard, went to Pennsylvania with Harold Denton.

At Gilinsky's urging, Ingram, a slight, unassuming man, opened two rooms on the fifth floor of the East-West Towers in Bethesda to the press. For technical information, engineers Brian Grimes and Dudley Thompson were stationed in the jury-rigged press center. They were given no training in how to deal with the press and no guidelines about what to say or not say. As a result, they speculated freely.

When the subject of the hydrogen bubble was raised by reporters in Bethesda, Grimes drew a schematic diagram of the reactor on a blackboard and explained the problem. When a reporter asked, "Is a meltdown possible?" Grimes simply said, "Yes, there is a possibility . . . I think everyone technical would have to concede that if the reactor was de-pressurized and the hydrogen bubble expanded, that there was a possibility—not a likelihood—but a possibility that things could go wrong, or all the cooling systems quit."

At 4:02 P.M. United Press International ran a story based on what Thompson and Grimes had said in Bethesda:

"Dudley Thompson, a senior official in the NRC Office of Inspection, said the threat is posed by a steam bubble inside the reactor that could increase in size as pressures in the reactor are lowered, leaving the core without vital cooling water.

"'We are faced with a decision within a few days, rather than hours, on how to cool down the core,' Thompson told reporters at NRC news center.

"'We face the ultimate risk of a meltdown, depending on the manner we cope with the problem. If there is even a small chance of meltdown, we will recommend precautionary evacuations.'"

Later Thompson said, "UPI just misquoted the hell out of me, and it got me in all sorts of hot water," but Jody Powell confirmed his statement at 5:15 P.M., saying, "The possibility of meltdown was mentioned. It is not an inaccurate report."

The UPI meltdown story first reached the television networks in the form of a bulletin sometime before 4 P.M. Jim Jensen of CBS went straight on the air with the story. So did Frank Reynolds for ABC. At NBC Edwin Newman was given the same job, but he hesitated.

"It just didn't sound right to me," Newman said. "All we had was a wire-service lead that was very frightening. I didn't think it should go out over the air without being checked."

Les Crystal, then president of NBC News, walked through the newsroom at this time and agreed. The story should be verified.

Alan Mohan, Newman's assistant for twenty years, called the NRC, trying to get in touch with Edson Case, as Harold Denton's second-in-command. Case's secretary would not put Mohan in contact with her boss. "She was only trying to do her job," Mohan recalled, "but I was trying to do mine too, and I needed to talk to that man. I tried to explain the problem, the urgency, and all she could tell me was that he was unavailable for comment."

With two minutes to go before breaking into the regular network program, Edwin Newman was already seated in the broadcast position. He took the phone from Mohan.

"Miss," Newman said, his voice rising, "I am about to tell millions of people that they are in grave personal danger. If this is not correct, only your boss can tell me the truth. You wouldn't want to stand in the way of his telling the truth, would you? *Would* you?"

"In twenty years of working together," Mohan said, "it was the first time I had ever seen Ed Newman lose control."

Newman's question to the secretary was treated in Bethesda like a threat. *Ed Newman of NBC says he will say there is a meltdown unless we talk to him,* was the way the story reached Case. He spoke to Newman instantly and said that the danger was insignificant. And that's what Newman told viewers of soap operas and game shows on

185

NBC that afternoon, in sharp contrast to the more volatile reports on the other networks.

Jonathon Ward of CBS was dispatching his three television crews from the studios of WHP-TV in Harrisburg when a call came in from CBS News in New York. The facilities at WHP were limited. It was a cramped small-market UHF station. In addition to the network, there were crews from Japanese and British television, KNXT from Los Angeles, as well as the normal WHP news operation that was devoted almost exclusively to covering TMI.

"What is a meltdown and how bad is it?" Ward was asked by assistant Northeast bureau chief Cindy Samuels. "We've got a story from the UPI wire saying there is a possibility of a meltdown at Three Mile Island."

"We hadn't heard anything about a meltdown up to this point," Ward said later. With its limited resources, WHP did not receive the UPI wire. "I tried to spell it out—that a meltdown is the worst thing that can happen to a nuclear reactor and that it can kill thousands at its worst."

"What do you need?" Ward was asked from New York. The producer asked for two more crews, an editor, dosimeters, Geiger counters, a helicopter, a radiologist—"Oh, and a bus, just in case we have to get out of here in a hurry."

"What about radiation suits?" asked New York.

That night, two news crews, an editor, dosimeters, Geiger counters and Harry Esterita flew into town. Esterita, soon to be known simply as "Radiation Harry," was CBS's $750-a-day expert on radiation. And out in front of WHP, a chartered Continental Trailways bus was parked, waiting to leave on a moment's notice. Three shifts of drivers took turns drinking coffee at the twenty-four-hour restaurant down the street from WHP. Inside the bus, yellow rubber antiradiation suits occupied each seat. Jeff Bitzer, a reporter for WHP, went through the bus that evening. "I counted the radiation suits," he said, "and then I counted the people from the network. There were no extras."

*　　*　　*

Dudley Thompson had been quoted as saying that the core might melt down, and Jody Powell had substantially supported him in that statement, but that didn't mean that everyone agreed about the danger. Gary Miller, who was still running Unit 2 for Met Ed, saw no prospect of a core melt and said so. In reaction to the UPI story and Powell's statement, Met Ed issued a release of its own at 5:30 P.M. saying, "The situation is exactly as it was at 3 P.M. The reactor remains stable. Reports of a meltdown are unfounded."

But the news was out, and the national press did its best to inform the nation about how serious the problem in Pennsylvania really was. Only, to many of the people at the site—even the reporters—the situation just didn't seem as bad as the NRC in Bethesda and Washington and the stories coming out of New York would have had the public believe.

"Good evening," Walter Cronkite said that evening from New York. "The world has never known a day quite like today. It faced the considerable uncertainties and dangers of the worst nuclear power plant accident of the atomic age. And the horror tonight is that it could get much worse. It is not an atomic explosion that is feared; the experts say that is impossible. But the specter was raised that perhaps the next most serious kind of nuclear catastrophe, a massive release of radioactivity [could occur]. The Nuclear Regulator—Regulatory Commission cited that possibility with an announcement that, while it is not likely, the potential is there for the ultimate risk of a meltdown at the Three Mile Island Atomic Power Plant outside Harrisburg, Pennsylvania . . ."

Richard Wagner, the CBS lead correspondent at Three Mile Island, was critical of his own network's show. "They gave the impression that Armageddon was at hand," Wagner said. "It was Walter's tone that particularly gave this impression. We were on the scene, in front of the milk bottles—as we called the cooling towers—and it wasn't that serious."

If the national press went overboard, hyping the danger of meltdown, the local press tended in the opposite direction.

The community turned to radio and television as its primary

sources of information. With WHP carrying all press conferences on both television and radio, local citizens had nearly as much information at their disposal as Civil Defense officials like Kevin Molloy did. Yet there were still questions.

"There were hundreds of them," Jeff Bitzer of WHP-TV recalled. "People would call in and ask 'Is it true that once I leave my house I won't be able to come back for a hundred years?' The community was really upset. We wanted them to be upset if there was a reason to be upset, but we didn't want to scare the shit out of them because you were being careless about what you were doing. Mostly what we did was tell people what was being said. They wanted us to make decisions for them: Should I leave my house? What should I do with my dog? If I leave now, will I be able to come back next Tuesday? Mostly we said that's a decision you'll have to make. Here's the information we have been given . . . We may have been too cool. We'll know when people start dropping over dead. We'll know that we should have said, 'Hey, you'd better get the hell out of here!' But we didn't."

"We had a lot of stories we could have gone with on Friday," said Saul Kohler, editor of the two Harrisburg papers. "We knew about the heavy car sales that day, the number of doctors and nurses leaving town and the money that was brought in to keep the banks from having to close. We didn't go with any of those stories. We tried to avoid panic."

Across the river at WCMB in Camp Hill, Steve Liddick was having problems balancing the news coverage of his station. "We would get these inflamed reports from the Associated Press out of Washington," Liddick said, "going on the air with anything and everything they could think of. I had to go on the air afterwards and say that the situation was not as bad as AP has said. Then we had a Paul Harvey commentary from ABC where, sitting in his safe little studio in Chicago, Harvey said that the problems at TMI were all over. I had to go right back on the air again, saying that while the situation was not as bad as the AP had said, it was not as good as Harvey said. I have no respect for that man at all."

*   *   *

By early evening, Jack Watson and Jody Powell had decided how to deal with the so-called "information problem"—the contradictory assessments offered by the agencies dealing with the accident. Their plan was simple: there should be only three sources of information. The NRC would handle all technical questions through Harold Denton, closing out any other NRC statements from Bethesda, Washington or Harrisburg. Any questions about emergency preparedness or evacuation would be answered by Thornburgh's office, with the ready assent of Paul Critchlow. All other information would be handled by Powell's people at the White House. But the system was not without its problems. Met Ed, for instance, continued to release statements to the press. And Eileen Shanahan, chief spokesperson for HEW, noted, "Powell took full responsibility for answering questions about what the federal agencies were doing, without setting up any system for *getting* information about what we were doing to *give* the press. It just shut off information on this end completely."

The NRC commissioners were in continuous session from Friday on, with the problem of the hydrogen bubble dominating discussion. And it was during one of those discussions that another potential problem with the reactor came to light.

GILINSKY: [to Hendrie] What is the problem about just leaving it [the reactor] the way it is? Is it the growth of the bubble or—I mean, it does seem to have cooled down a bit . . .

HENDRIE: Yeah. I think it ought to stay the way it is probably for a couple of days. Over the long term, this is an unsatisfactory configuration for the machine to be in. We've got to get it cooled down.

GILINSKY: I mean, it's not moving at any significant rate the way the temperature is. Is that what you're saying?

HENDRIE: Well . . .

GILINSKY: Why wouldn't it go cold this way?

As the only one of the five NRC commissioners with a technical background in nuclear power, Hendrie had to interpret the accident for his fellows. As each new condition arose in the crippled reactor, he explained it to the others, that they might better judge what the NRC should do in response. Not that the commissioners even had a great deal to do with the NRC response; that was being handled by Denton and the others in Bethesda. But now, for the first time, Hendrie's thinking about the progress of the accident and of the hydrogen bubble had surpassed the present. Rather than reacting to the current situation, the NRC chairman was worried about a new condition that might come about, and he was the only one to feel this concern.

Hendrie told Gilinsky that they could not simply wait for the reactor to cool down as it was because the high radiation in the reactor might be acting on the water in the core, breaking it down into hydrogen and oxygen. He knew that if the concentration of oxygen in the bubble went over 4 percent, there was the possibility of an explosion. With the brittle, damaged core, an explosion could cause serious problems even if it didn't damage the pressure vessel. Just waiting as Gilinsky had suggested would take weeks or months before the reactor was in cold shutdown. The bubble couldn't wait that long. It might be only days before enough oxygen would be generated to cause a problem.

The NRC chairman called Roger Mattson to arrange for calculation of oxygen generation in the reactor:

HENDRIE: It's trivial, but I'm worried about the oxygen build-up.

MATTSON: Oxygen build-up where?

HENDRIE: In the bubble over there in the dome. Why do people—Why am I the only one worried about that? Nobody else seems to be.

MATTSON: That's the first time I've heard the question asked. I'm not an expert on that. Tell me why that's important.

HENDRIE: If you build up the oxygen content in that bubble, you get flammable.

MATTSON: Inside that—

HENDRIE: Yep. Can we stand the bang in there? That core doesn't sound to me like it's in much shape to get rattled.

MATTSON: We'd better get somebody thinking about that.

HENDRIE: And furthermore, the time to go before it gets flammable doesn't sound that far away. I had kind of hoped it would be a couple of weeks before we got to that stage . . . We may get pressed before that.

Harold Denton's afternoon and early evening at TMI had been occupied with learning the condition of the reactor and of the utility's response to the accident. With a little more information, Denton had changed his mind from the morning and no longer felt that a precautionary evacuation was in order. But his confidence in Met Ed was less than total.

"I guess I've developed a management concern about the capability of the utility here to cope with new problems that come up," Denton told the commissioners in a telephone conversation. "They're stretched very thin in some areas. I've discussed it with the local management and with the management of GPU. I think they need stem-to-stern reinforcements down here in many areas."

Denton decided to call for help from experts throughout the nuclear industry. If Met Ed didn't want to admit they needed help, then Denton would admit it for them. He called his boss, Jimmy Carter. As the personal representative of the President, the nuclear engineer was learning to use the power that had been given to him so arbitrarily on Friday morning. Carter told Jack Watson to do whatever was necessary. Watson called Herman Dieckamp, president of General Public Utilities.

"I underscored the sense of urgency that Harold Denton felt and asked for the company's cooperation in getting those people assembled as quickly as possible," Watson recalled. "Mr. Dieckamp pledged his fullest support for his company to get that done. And, in fact, it was done quickly."

Joe Hendrie described the call, in rather different terms, as "Jack Watson reading him the riot act."

It was Thornburgh's false impression that Denton had been sent by Carter solely to keep the governor's office informed of conditions at the plant, and Denton did brief state officials at 8:30 P.M. He relayed his new feeling that an evacuation was not called for at that time, but he also tried to explain just how badly damaged the reactor core was—that at least a third of it had been exposed, some melting had taken place and any large shock to the brittle remains of the core could send the twisted fuel rods crashing into one another, concentrating the effects of remaining decay heat and making a safe shutdown even more difficult. And he told Thornburgh about the bubble of hydrogen, floating above the cooling water at the top of the reactor pressure vessel. By now there was an estimate that the bubble occupied one thousand cubic feet of the 15-by-40-foot cylinder. That was one eighth of the total volume—and at a pressure of more than a thousand pounds per square inch. Denton told the governor that if anything should happen to drop pressure in the cooling system, that bubble could grow and fill the vessel, pushing out water and uncovering the entire core. A meltdown was remote—but possible. What was missing from Denton's account was Hendrie's concern about the bubble's exploding. But isolated from Bethesda, Denton still did not know of Hendrie's latest worry.

At 10 P.M. Thornburgh and Denton met the press in the cramped capitol media center. It was the last time the small room would be used for such press conferences. It had simply been outgrown by Three Mile Island.

After Thornburgh finished a short statement reaffirming his advisory that pregnant women and preschool children remain outside the five-mile radius, but saying that his other advisory for citizens to remain indoors would expire at midnight, the assembled press leaped forward with their technical questions. Thornburgh turned to Denton.

"I was standing there, backing up the governor, looking at my shoes a lot," Denton recalled later, "and suddenly I was in front and they were asking me questions."

It was Denton's first substantial performance before the press, and he did not disappoint those who had sent him. Denton, who was not only careful and precise, but also generally honest, told the reporters, "This is easily the most serious situation in the life of the reactor program."

Under the sort of "what-if" questioning that the press loves and engineers hate, Denton was asked whether the bubble could explode. Denton said, Yes, it could explode, but only if there were enough oxygen and a spark, and he didn't think there was likely to be either in the humid, hydrogen-rich environment of the reactor vessel.

Saturday morning, going for sensation as usual, the New York *Daily News* was the only major newspaper to lead its TMI story with the possibility of the bubble's exploding. In a headlong rush for high circulation, the editors of the New York tabloid didn't know they had published NRC Chairman Joseph Hendrie's greatest fear.

18

Potassium Iodide
and
Submarine Sandwiches

The Incident Response Center in Bethesda had gone on twelve-hour shifts by Friday night. Roger Mattson and his team of engineers were hard at work calculating just how much oxygen was accumulating in the bubble. A small staff of briefers called influential members of Congress to keep them informed of new developments in the accident. Several more men relayed information back and forth from Harrisburg and H Street. Others called experts throughout the nation asking for advice or scoured military and GSA warehouses for this bit of hardware or that. They were still looking for a robot to send into the lethal radiation of the containment—the carcass of one such machine was already sitting where it had dropped dead in the high radiation that afternoon.

Bernie Weiss, who seemed to be always on the phone that night, serving in a variety of capacities, was having his own acquisition problems. "I'm ordering seventy submarine sandwiches and you say you don't deliver!" he screamed into the phone.

Not all calls coming into Bethesda were from engineers, reporters or politicians. "I've got a very angry man on the other line," the NRC operator said. "He says that he's a truck driver and he spent last night near Harrisburg and now they won't accept his load of lettuce at the produce terminal in New York. He thinks the NRC should pay for his lettuce. Who should I have him talk to?"

Every other hour, a call would come in from the office of California Governor Jerry Brown. Brown was worried about the Rancho

Seco nuclear power station near Sacramento—a twin to TMI Unit 2. He wanted the NRC to shut it down.

"This guy's trying to ride TMI into the White House," one of the Bethesda engineers said after the third call from Brown's office.

In the hours between Brown's messages, frantic appeals were coming in from the California nuclear industry. "You can't do it!" one executive urged. "Brown is trying to put us all out of business and you can't let him. There is no reason to shut any plant in this state. It's all politics."

Dudley Thompson left Bethesda's small press center for a few hours to look for a hundred tons or so of lead bars. If there was oxygen in the bubble, then one possible solution would be to use the hydrogen recombiners—units attached to the reactor vessel for making hydrogen and oxygen into water. But only one of the two recombiners was working, and that one leaked. If they had to use it, the unit would have to be shielded with lead.

"How much do we need?" Thompson asked someone at TMI.

"Anything you can get your hands on. We'll need tons of it, in bricks."

By late morning on Saturday, 150 tons of lead bricks were being piled around the recombiners of Unit 2. Some of the bricks came in by truck, but most—at a dollar a pound—flew in by helicopter.

Following Jack Watson's organizational meeting at the White House on Friday, Bob Adamcik of the Federal Disaster Assistance Administration was sent to Harrisburg as the federal disaster coordinator. It was his job to help PEMA and the affected counties develop evacuation plans for five-, ten- and twenty-mile distances from TMI.

The twenty-mile radius was a new requirement urged by Dr. Arthur Upton of HEW's National Cancer Institute. The area included more than 675,000 people, 13 hospitals, 20 nursing homes and a prison. Three hours is all the disaster planning teams under Adamcik decided it should take to evacuate the large area. "I'm glad we never had to test that estimate," one PEMA staffer said later.

When not enough staff arrived on Friday night to run the Frey Village Nursing Home in Middletown, Kevin Molloy had the residents evacuated—a job that alone took much more than the three hours alloted for evacuating the whole county.

Nobody got less sleep on Friday night than Jerry Halperin of the U.S. Food and Drug Administration. He spent the night phoning all over the country, looking for a million doses of potassium iodide, one of the few drugs that can help prevent radiation poisoning. And a drug that was not available in any quantity.

A continuing concern of health professionals dealing with the problems of Three Mile Island was the presence or absence of radioactive iodine. It was constantly expected to appear in water, grass and milk samples. It was the major stated reason for evacuating pregnant women and young children.

The human body uses small amounts of iodine in the thyroid gland, and when the iodine taken up by the thyroid from food or water is radioactive, it almost always results in benign thyroid nodules or in thyroid cancer. It doesn't take much iodine to do the job, and those exposed may not notice anything for years, but it will catch up with them eventually.

In England, where large amounts of milk that was high in radioactive iodine had to be impounded after the Windscale reactor accident of 1957, a medical treatment was developed to prevent thyroid poisoning by radioactive iodine. It was discovered that if those likely to be exposed to radioactive iodine 131 consumed a drug rich in iodine *before* being exposed to the radioactive form, their thyroids would be so "filled-up" with iodine that the radioactive variety would be rejected. That is why packets of a similar antiradiation drug—potassium iodate—are stored in electric meter boxes of homes around nuclear power plants in England. The drug, which has a short storage life, is changed regularly by meter readers and is available for residents to use on instructions of local officials.

In mid-1977, a report was issued by the National Council on Radiation Protection and Measurement, urging use of potassium iodide to protect workers and general populations during a radiological emergency. Testing by the Food and Drug Administration's Bureau of Drugs followed and approval of the new/old drug came

196

in December 1978, when manufacturers were told they could go ahead and produce potassium iodide pills or liquid.

When the accident at Three Mile Island came along, no drug company had taken the FDA up on its offer. There just didn't seem to be a market for the drug. It had a relatively short shelf-life, the government didn't plan to buy supplies for protection of citizens during World War III and the only other medical use potassium iodide has is as an expectorant in the treatment of asthma.

On Friday Dr. Donald Fredrickson, Director of the National Institutes of Health, recommended that potassium iodide be administered to workers on-site as soon as possible. John Villforth of the FDA's Bureau of Radiological Health and HEW's Three Mile Island coordinator, sent on the recommendation to both the NRC and to Tom Gerusky at BRP.

Gerusky agreed that it would be nice to have potassium iodide available and that the state would accept it if Villforth could find any of the drug. That's when Villforth turned to Jerry Halperin, assistant director of the Bureau of Drugs and a former pharmacist, to find adequate supplies within a few hours.

"We first thought there might be enough of the stuff sitting on pharmacy shelves in Pennsylvania," Halperin remembered. "Most pharmacies will keep a small amount on hand for prescription use to treat asthma. But then I started calculating how much we would need to treat the [local] population, and it came out to somewhere over a ton of crystalline potassium iodide. I stopped looking on pharmacy shelves."

Checking reference books and calling drug companies on the East Coast, Halperin soon learned that no company had a ton of potassium iodide available.

"There are two forms of the stuff," Halperin said. "There is a chemical and then there is a drug. If I couldn't find a drug company, I started to look for a chemical company with a pharmaceutical division that might be able to make the chemical into a drug."

At 3 A.M., Jerry Halperin found Mallinckrodt, Inc., in St. Louis, a chemical company with a pharmaceutical division. But Mallinckrodt, Inc., was closed for the night. Eventually a night security

197

guard answered the telephone and Halperin tried to explain his purpose for calling at this unseemly hour. A very persuasive bureaucrat, he finally got the home telephone number of the company president.

Halperin quickly arranged with the sleepy chemical-company executive to take more than a ton of potassium iodide, place it in solution and fill a quarter million one-ounce bottles—and it had to be done that weekend. With nothing in writing at all, $400,000 dollars of government money was committed, and production of the drug commenced. But there were still problems. Mallinckrodt didn't have any medicine droppers to fit the 250,000 bottles they would be filling in a few hours. And the drug could not be accurately dispensed without medicine droppers.

Halperin began waking up businessmen looking for 250,000 medicine droppers. He eventually found a supply in New Jersey, but the dropper caps were too big to fit the bottles. Two hundred fifty thousand rubber bands were ordered to attach droppers to bottles.

Mallinckrodt wasn't sure it could make and bottle all the drug in time, so Halperin went back on the phone, eventually getting Parke Davis and Co. in Detroit to fill 97,000 bottles with the Mallinckrodt solution. And Parke Davis even had bottles with droppers!

The first bottles left for Pennsylvania on an Air Force C-130 transport Saturday night. There were no droppers, the small brown bottles were unlabeled—in direct violation of federal regulations—and many of them had been overfilled so that potassium iodide residue crystallized on the outside. "The first batch didn't look too good," Halperin said, "but they would have done the job."

Mallinckrodt and Parke Davis would continue to work until 237,013 bottles were delivered by April 4. The Air Force would only deliver the first batch of 25,000 bottles. Since President Carter had not declared a disaster or state of emergency, the Air Force was not assured of being paid for its services and so refused to carry any more of the drug to Pennsylvania. It would take several months for Mallinckrodt, Inc., to receive its $400,000 too. And Parke Davis never even sent a bill. Fearful of product liability suits stemming from bottling another company's drug, the Michigan pharmaceutical maker chose to donate its bottles, droppers and labor to the cause.

Late into the night, Roger Mattson and his group of experts continued to compare calculations of oxygen evolution in the reactor. Bethesda was as active at midnight as it had been at noon.

In the picnic area surrounding Three Mile Island's observation center, trailers were being set up by Met Ed and the federal agencies. Soon to be called Trailer City, it was the new center of activity for the NRC, Met Ed and the press at TMI. With more than forty trailers dragged from construction sites or rented for the occasion, offices, showers and mess trailers were set up to take care of the more than five hundred technicians and engineers who were on the job around the clock. A line of chemical toilets replaced the facilities lost on the Island when Gary Miller was told to stop dumping wastewater from the flush toilets. Typewriters and Xerox machines ran constantly, an indispensable part of dealing with nuclear accidents.

Fifty yards away, on the other side of route 441, between the observation center and the river, CBS had set up a huge Winnebago motor home with a color television camera on its roof, trained on the power plant. The Winnebago dealer in Middletown who'd rented it to Jon Ward said the network would have to buy the $28,000 unit if it became contaminated with radioactivity. New York said "Sure."

Technicians and reporters weren't the only ones awake on that night in Pennsylvania. Mrs. Ben Ebersole of Londonderry Township answered her telephone at 2 A.M. It was a stranger calling from Frankfurt, Germany. He asked, "Are you the lady who owns Three Mile Island?" She said that she did not own the Island but did live only about a mile away. "If you don't own it and you live so close, why are you still there?" the caller asked.

Saul Kohler of the Harrisburg *Patriot* and *Evening News* went to see the 2 A.M. showing of *The China Syndrome* at the Camp Hill Cinema across the river from Harrisburg. "The show was packed," he recalled. "And afterwards we went to Howard Johnson's for a cup of coffee. Three-thirty A.M. and there wasn't an empty seat in the place."

About 150 people—families, pregnant women and young children —were trying to sleep in their cardboard bedrooms on the floor of the Hershey Arena. The Red Cross had been able to set up its shelter on relatively short notice, but it was not prepared to deal with the overwhelming media interest that extended into the night.

"They just invaded," Red Cross director Edward Koast recalled. "We had more media people than we had shelter occupants. We had teams and teams of camera crews all over the place, poking cameras in people's faces while they're sleeping or while they're feeding the kids. Throwing questions, all kinds of questions at them from all angles, everybody was trying to get the attention of the evacuees . . . Charter planes were landing in Hershey, charter helicopters were landing right alongside the arena up there, and news crews were running into the place . . . People who looked more distraught were getting more attention than those who seemed to be relaxed and not as distraught . . ."

But the press was not without concern for its own safety. John Baer of WITF-TV in Hershey recalled, "I shipped our two small children and my wife to Washington on Friday. By the time I got here, they were talking about a meltdown. When they said 'possible,' I thought 'probable' and I got scared. Soon there was hardly anyone still working here. Mostly the men were under incredible pressure from their wives. We have a woman producer who is married to a doctor, but she was never under any real pressure from him. We interviewed Nina Tottenberg of National Public Radio on one of our shows. She said something that really hit home with me. She said that she knew reporters who had spent a year in Saigon and had turned down this assignment."

"I put a lot of false hopes in my dosimeter," said Jeff Bitzer of WHP-TV. "Intellectually you knew that by the time the dosimeter said anything it was too late anyway, but I went to bed that night with my dosimeter on the pillow and I clipped it on my bathrobe in the morning. I live seven miles from the plant and I sent the people I live with out of town right away. In fact, I spent one night at my parents' house with my dog. But I got concerned at some point about looting, so I just decided to stick around. You got pretty fatalistic about it. It was like it was too late anyway. If you're hit, you're hit. But if you've got to go, it sure beats getting run over by a truck. At least you'd die doing your job. I looked around the house and tried to decide what I would take with me in the event of an evacuation. You know, there wasn't anything I'd take. That was strange."

19

More Information Problems

Saturday morning Tom Gerusky met with State Health Secretary MacLeod and other state officials to inform them of the impending arrival of the potassium iodide. Up to this point, MacLeod maintained that he had heard "something" about the use of potassium iodide on the previous day, but had not been aware that supplies were available. Nor did he know the technical details of administering the drug. Nevertheless, MacLeod accepted reponsibility for the potassium iodide from Gerusky.

Eventually MacLeod decided not to distribute the drug at all. He thought the public might take overdoses, or might properly use the drug and then refuse to evacuate, thinking that they were protected from all ill effects of radiation. Distribution of the drug would emphasize the danger to public health posed by the accident and might make the public feel that the governor was lying when he said there was little to fear. Finally he decided not even to allow the drug to be placed in police or fire stations for distribution at the last minute —the stations might be stormed by citizens demanding medication.

Jerry Halperin's all-night effort and several hundred thousand government dollars eventually resulted in 237,013 bottles of potassium iodide being held in a secret state warehouse under armed guard. "It was hotter than gasoline," John Villforth said. But HEW Secretary Joseph Califano, his agency having gone to considerable effort to find and produce the drug, was not about to have it placed in storage by the state. In a series of memos from Califano, his deputy

Rick Cotton and Dr. Arthur Upton of the National Cancer Institute, the White House was constantly called on to force distribution of the drug or at least announce that it was available. Watson would have none of Califano's suggestions.

Despite MacLeod's attempts at secrecy, word of the potassium iodide got out to the local press. Mary Bradley of the Harrisburg *Evening News* made calls to St. Louis and found out about its production, talked to the Air Force pilot who delivered the first shipment, and then she hit a brick wall. "It was like the stuff came into town and then just disappeared," she said. "I knew what it was, who had made it, everything but where it was, and they wouldn't tell me."

Jeff Bitzer also knew that something important had come in on an Air Force plane, but he didn't even know what it was. Bitzer called the state General Services Department, which he knew would have stored the drug shipment on arrival. "You know that stuff that came in by Air Force plane last night?" Bitzer asked. "Was that in warehouse three or warehouse four?" Bitzer's contact in the agency said, "No, dummy, it's in warehouse nine." "When they won't tell you anything," Bitzer recalled later, "they'll always tell you if you're wrong."

Bitzer took this new information to the state Drug and Cosmetics Bureau, saying, "We're not asking if it's here, we don't want to know if you're going to distribute it, all we want to know is what it does." But Paul Critchlow had effectively blocked the release of any information about the potassium iodide shipments to Pennsylvania. Frustrated, Bitzer eventually went to Critchlow for answers. "Look," Bitzer said, "we know where the stuff is being stored and we'll go on the air with the location unless you guys cooperate and talk to us about it."

Two days later the press was allowed into a guarded state warehouse near the airport to photograph crates of potassium iodide. Fifteen minutes after the photographers left, trucks and fork lifts drove in to move the drug to a new and secret hiding place.

From Saturday morning on, the problem of technical information about the accident and its seriousness became even more profound. As the NRC experts in Bethesda became increasingly uneasy about the size of the hydrogen bubble and its oxygen content, Gary Miller and the Met Ed operators actually at TMI were becoming increasingly sure that they had the accident under control. But Met Ed, through its overoptimistic statements of Wednesday and Thursday, had destroyed its credibility with the public, press and even with the NRC, so Miller and Herbein were just not believed when they said that the bubble was getting smaller or that its oxygen content was not an important issue. Even the NRC forces in Pennsylvania, first under Gallina and then under Denton and Stello, were suspicious of Met Ed's findings. But they were also learning not to trust everything told them from Bethesda, either.

Met Ed had been told by the White House to stop making any comments to the press, but a press conference was already scheduled for 11 A.M. Creitz and Herbein decided to make it their last. And they had good news: The bubble was smaller and shrinking by the minute.

Creitz announced that there would be no more press conferences for Met Ed and then turned the podium in the Legion Hall over to Herbein, who claimed that the bubble had gone down in size from 1,000 to 800 cubic feet. "I personally feel the crisis, if that is what you want to call it, is over," said Herbein.

As for the likelihood of an explosion in the reactor building, Herbein said, "I think that that potential exists, but I think it is exceptionally minimal."

As Jack Herbein was taking heat from the press in Middletown, Gary Sanborn was making final contact with an engineer in Met Ed's Reading headquarters. While a lot of the press was looking toward the resolution of the accident or concentrating on the hydrogen bubble, Sanborn was still trying to understand what had gone wrong in the first place. He asked his source about why the auxiliary feedwater pumps had not been able to feed water to the steam generators for several minutes on Wednesday morning. "The valves were closed," his source explained. Were they supposed to be closed? No.

"We had the first indication of operator error at that point," Sanborn claimed with glee. In its Sunday edition, the Allentown *Call-Chronicle* would be the first newspaper in America to say that operator error had contributed to the problems at Three Mile Island.

But few people read the Allentown *Call-Chronicle*. It was Tuesday before a similar story appeared in the Chicago *Tribune*. The Associated Press would pick up that story, attributing it to the *Tribune* until Ben Livingood called to set them straight.

An hour after Herbein and Creitz made their avowed final statement to the press, Denton faced the same reporters across town in the auditorium of the Middletown Borough Hall. Karl Abraham had set up operations in the kitchen of the hall—the largest public building in town and the site of NRC press conferences for the next several days. On Monday, the senior citizens' hot-lunch program would serve dessert from the top of Abraham's Xerox machine.

Denton contradicted much of what Herbein had said an hour before. No, the bubble did not seem to be shrinking. The crisis was not over and would not be over until the reactor was in a state of cold shutdown. For the fourth day, core temperatures remained at 280 degrees. That was a long way from cold shutdown.

While Denton was talking to the press in Middletown, Mattson heard from his consultants that there was 2 to 3 percent oxygen in the bubble and it would be 4 to 5 days before combustion could take place. But late in the afternoon, Robert Tedesco of Mattson's staff reported back with other calculations made by the Knolls Atomic Power Laboratory indicating that the bubble appeared to be approaching the threshold of flammability, although spontaneous ignition seemed unlikely. It could be just a matter of hours before an explosion was possible.

With conflicting calculations, all of them looking bad, Joseph Hendrie, chairman of the NRC, decided he should speak to the press himself. Hendrie, like Thornburgh, Denton and even President Carter, was caught up in the problems of informing the press and the public. He and the other commissioners spent hours laboring over amateur press releases designed to tell enough, without saying too much. Each time new, conflicting information came out of his agency, Hendrie felt that he could have, and should have, presented it more clearly himself. So, in direct violation of the directive laid down by Jody Powell the night before, he spoke to the press on Saturday afternoon—with very mixed results.

When asked whether the hydrogen bubble could explode, Hendrie replied, "As long as the bubble has a hydrogen steam fission product composition, why, it's not flammable. But if enough oxygen over a long period of time were evolved, why, it could become a flammable mixture." Hendrie did not say that this is what he thought was taking place, that the bubble might explode. But it was.

Richard Thornburgh, his wife Ginny, William Scranton and his wife Coral made an appearance at the Hershey evacuation center on Saturday afternoon. There wasn't much to say. The 160-odd people sitting on folding chairs looked at the governor and he looked back. Children were running everywhere, a band was playing and a television was on in each corner of the arena. Saturday had at least provided the children with cartoons. Coral Scranton, obviously pregnant, talked quietly with a group of frightened pregnant women as television cameramen tried to shoot the scene without being noticed. Thornburgh shook hands, made small talk and tried to avoid the constant question—"When can we go home?" There was no answer to that question: nothing was better; nothing was worse. Pictures taken, smiles gone, Thornburgh was back at work by 3:45 P.M., just in time to hear from Harold Denton, who called to say something about how there just might be some oxygen in with that hydrogen bubble and it could reach the point of flammability.

Oran Henderson and Kevin Molloy were both beginning to know frustration on Saturday afternoon. Henderson had been the major source of information for the Civil Defense system since Wednesday, but he was no longer being invited to the inner briefings with Denton, Scranton and Thornburgh. Henderson had been too eager to take command of the situation as a military leader. He was ready to move, ready to lead an evacuation, and did not have, in the eyes of the politicians above him, a clear view of the political nature of the situation, the fact that moving people from their homes without absolutely clear cause would be political suicide—worse than not moving people at all, even if the danger was real. So, when it became evident that Henderson could not be trusted to keep quiet and follow

206

the governor's policy in what was Henderson's first real action since Vietnam, the colonel was sent back to his office to update evacuation maps and wait. "The lieutenant governor thought I could best serve back at PEMA headquarters," the old soldier said.

With Henderson out of the flow of information, Molloy felt particularly vulnerable. He had spent much of the night before coordinating evacuation of invalids and old people from the Frey Village Nursing Home in Middletown. Now Molloy was hearing on the news that there might have to be a twenty-mile evacuation and that the feds thought it could be done in three hours. Charged with the safety of more than 100,000 people, yet having to rely on WHP radio to tell him what was happening at TMI, Kevin Molloy was beginning to feel like a tool, used by the state, a man trying to do his job blindfolded.

In Washington, Richard Cotton of HEW was angry too. "Do you honestly believe this?" he asked Jack Watson after reading the estimate that nearly 700,000 people could be moved in three hours. Watson didn't even answer; he, too, was mad—though this time at Cotton and his boss, Joe Califano.

HEW was feeling frustrated by Jody Powell's public information guidelines. Production of potassium iodide had been ordered by HEW with the expectation that it would be put to some use in Pennsylvania, but Powell's gag order on the federal agencies was blocking distribution of information on how to use the drug if it had to be given to the public. Late the night before, Eileen Shanahan, chief spokesperson for HEW, had gone "screaming and hollering" to Cotton for some action. "We have important information that the people of Pennsylvania should know," Shanahan pleaded, "and we have no way of getting it out."

Cotton asked for at least a joint press conference with the White House. Watson said no.

"We didn't need what was obviously just going to be a Califano media event," said a White House staffer after the meeting.

But there was more. Cotton relayed the feelings of experts in his department that a precautionary evacuation was in order, that potassium iodide should be distributed to the workers on-site and a distribution plan devised to get it to the general public within a few hours. "That's the governor's business," Watson said.

"Then let us talk to the governor," Cotton asked.

"Write a memo," Watson replied.

Potassium iodide was of concern to the NRC at Three Mile Island, too. They wanted some on hand for their own use just in case of an accident. Health Secretary MacLeod said no. "If I give it to you, I'll have to make it generally available." Even HEW, which had ordered the stuff in the first place, was denied access to the drug. A Public Health Service physician wrote a prescription instead, and it was filled at the local Giant Pharmacy.

The press and people of Pennsylvania were getting ready for what they would later call "fright night." Four-year-old Brandy Straub told reporters about her dream of the night before: "It was a big ball, and you know the way things glow? It glowed like that. And then there was a witch and the big ball killed everybody. And all the cats and dogs and rabbits were dead."

A page in the state legislature, around thirteen years old, with blond hair and freckles, looked out a large window in the capitol toward the Susquehanna: "Look—coming out the river—it's the Amazing Hulk!"

Jimmy Breslin, who had set up operations in Santana's, favorite bar of the Harrisburg press, walked out of the men's room. "Something's happened," he said. "It came off in my hand."

At 8:27 P.M. an editor's advisory went out over the AP wire saying that a story would soon follow on how the bubble might explode. Reporters leaning over the wire machine in the capitol newsroom stood up after reading the advisory and marched into the governor's office demanding the truth. They were scared.

Word of the advisory reached Casey Bukro of the Chicago *Tribune* at Santana's, where he and half the press corps were eating. Bukro confirmed the story with an NRC source in Washington and the press began to stream out of the restaurant, many of the reporters stuffing food in their pockets as they went.

Ellen Hume of the Los Angeles *Times* turned to Bukro and asked, "Does this mean we're all going to die?" He did not say no.

Bukro and the entire *Tribune* contingent moved to new quarters seventy-five miles from Harrisburg, "interviewing evacuees on the way."

Dick Lyons of the *New York Times* called a number given to him by Joe Fouchard, chief public information officer of the NRC. It was the super-secret number of Harold Denton's hotel room. Denton said he would come to the capitol newsroom to answer questions.

NBC ordered a plane to stand by at the airport in case a quick getaway was required. Jonathon Ward of CBS noted with some glee that the evacuation plans would shut off traffic to the airport, stranding his competition. The CBS bus was more practical. Ward made plans to remain with a cameraman and a helicopter to catch as much of any last moments as he could. Rod Nordland and Joel Shurkin made similar plans for the Philadelphia *Inquirer.*

Jim Panyard of the Philadelphia *Bulletin* called his editor and asked what to do in case there was an evacuation. "They thought for a moment and said, 'Well, what do you think you should do?' " Panyard recalled. "I said I thought I should get my ass out."

The AP editor's advisory that created so much concern in Harrisburg and Washington was followed at 9:02 P.M. by a story written by Stan Benjamin:

> HARRISBURG, PA. (AP)—FEDERAL OFFICIALS SAID SATURDAY NIGHT that the gas bubble inside the crippled nuclear reactor at Three Mile Island is showing signs of becoming potentially explosive, complicating decisions on whether to mount risky operations to remove the gas.
>
> Officials said earlier that tens of thousands of people might have to be evacuated if engineers decided to try to remove the bubble, operations that could risk a meltdown of the reactor and the release of highly radioactive material into the atmosphere.
>
> But the Nuclear Regulatory Commission said Saturday night that it might be equally risky not to try the operation, because the bubble

209

showed signs of gradually turning into a potentially explosive mixture that could wreck the already damaged reactor . . .

Benjamin's story was not the product of any inside information. Edson Case and Frank Ingram of the NRC had told him that the bubble might explode. He asked them and they answered—it was what they thought might happen. Ingram even called Benjamin back to check the story before it ran on the wire, confirming that it was accurate. Then Benjamin added Hendrie's speculation from the afternoon press conference to round out the story. It was all factual in the eyes of the NRC in Bethesda and all wrong to those at Three Mile Island. "I was never worried about an explosion in the bubble," said Met Ed's Gary Miller. "No way could it blow." With the torturous NRC communication setup, Miller never spoke to Bethesda, either. "There was no oxygen in the bubble. We didn't realize the concern that existed in Bethesda."

The high radiation that exists in the core of a pressurized water reactor like Unit 2 causes almost constant dissociation of water into hydrogen and oxygen. Much of the oxygen released is swept out of the core after it has been dissolved in the primary coolant, but there is enough concern about gas voids and explosive mixtures that another safety system is also used to combat free oxygen. Extra hydrogen gas is dissolved in the coolant, causing an overpressure of hydrogen that results in the spontaneous recombination of most of the oxygen with hydrogen to form water. Vollmer and Stello in Pennsylvania knew this. In fact, at one point on Friday the press in Pennsylvania was told the bubble had been formed from this dissolved hydrogen.

But Roger Mattson didn't know about such things as dissolved hydrogen or overpressures. As he had admitted to Hendrie, "I'm not an expert on that." Overlooking the fact that seemed so obvious at Three Mile Island, Bethesda had gone berserk. Earlier in the evening, Stello had called to argue the point with Edson Case:

"Look at it this way," Case had said to Stello, "you've got six or seven guys out there with slide rules. Mattson has Tedesco and maybe fifty other guys working on the problem. You're outnumbered maybe five to one. Now, who's right?"

Soon after the AP story ran, Harvey Tate, the manager of WCMB radio in Camp Hill, called Bethesda in desperation. "If you're going to deny the AP story, you'd better do it now," he warned, "because things are going crazy here in Pennsylvania."

A panic-stricken woman called Dauphin County Control—Kevin Molloy's operation—with disturbing news. "I heard it," she shouted. "I heard the bubble explode!" Then she heard the noise again as a second C-130 transport plane screamed low over her house on its way into the Capitol City Airport with a load of lead.

Paul Critchlow hurriedly wrote a press release at 9:30 P.M. after speaking to Denton on the telephone:

> HARRISBURG (MARCH 31)—THE NEWS REPORT THAT THE GAS bubble in the nuclear reactor is becoming potentially explosive is not true, according to Harold Denton, director of the Office of Nuclear Reactor Regulation . . .

When word of Stan Benjamin's story about the explosive bubble reached the NRC trailer at TMI where he was working, Vic Stello could not stand it anymore. He called Bethesda to find out what had gone wrong:

STELLO: Have you got a copy of the AP story?

CASE: We—the AP guy read his lead to us. It sounded all right. I—I don't know what I can do about it, Vic. The Chairman did say, and I did say, that there's a possibility of explosion in the reactor pressure vessel. It would be caused by hydrogen and oxygen combining; the oxygen is continuing to be evolved through radiolysis; we're calculating the rate; and this is the risk of continued operation in this mode. All facts . . .

STELLO: Well, then you're telling me that the AP story is quoting what you said?

CASE: Yeah.

STELLO: Huh?

CASE: Yeah.

STELLO: Didn't you think it's unfactual?

211

While Stello was talking to Case, Joe Fouchard was talking to Frank Ingram, his assistant, who had listened to Benjamin's story and okayed it.

> FOUCHARD: . . . We've got a serious situation up here. People are leaving hotels and stuff like that.
> INGRAM: Yes, I hear.
> FOUCHARD: No more statements, you know, from down there.
> INGRAM: Uh-huh.
> FOUCHARD: What the hell happened down there . . .?
> INGRAM: . . . Well, apparently they moved an urgent flash that NRC was now saying that it was dangerous. We heard that Stan had moved it as an advisory and we called him on it . . .
> FOUCHARD: Did you kill it then, or what?
> INGRAM: Pardon me?
> FOUCHARD: What did you do about it then?
> INGRAM: It was factual.
> FOUCHARD: What do you mean it was factual!
> INGRAM: What Ed and the Chairman had said.
> FOUCHARD: What did you say? You didn't say there was any imminent danger, did you?
> INGRAM: We said there was a potential explosion in that bubble.
> FOUCHARD: Who said that?
> INGRAM: Ed said it, I said it, and the Chairman said it.
> FOUCHARD: What?
> INGRAM: That the amount of oxygen was increasing very slowly and it was a matter of X days away before it would reach that point . . . I don't think there's anything that we've said that's inaccurate—I'm sorry about that.
> FOUCHARD: Don't get upset.
> INGRAM: Well, I am upset. I feel like I've botched the whole thing . . .

As the furor in Pennsylvania was felt further up in the federal establishment, the decision was made by Jack Watson that there would be no more statements to the press by Ingram or anyone else in Bethesda, and no more statements by the Chairman of the NRC. Denton was the sole source and his credibility had to be "sustained and nurtured." At 11 P.M. the NRC in Bethesda stopped answering calls on their press lines. They just let the phones ring.

Just as Stello and Fouchard did not share Bethesda's concern about the bubble, some of the reporters in Pennsylvania doubted the AP story too; they tried to keep their publications from carrying Benjamin's piece in the Sunday editions. Ben Livingood, who had a pretty good idea of what was taking place on the technical front, called just in time to stop his editor from remaking the next edition's front page, headlining the AP story. Joel Shurkin killed the AP story at the Philadelphia *Inquirer;* Dick Lyons got it dropped by the second edition at the *New York Times.* But most of the nation would awake the next morning to Sunday papers featuring gloom and doom and hydrogen explosions. Many readers, and even a few editors, came to confuse "H-blast"—the media shorthand for a potential explosion of the bubble—with what happens in a hydrogen bomb.

At 11 P.M., Harold Denton, tired, needing a shave, walked into the capitol newsroom. He was, as usual, a calming influence on the press. Denton did not deny that the bubble might eventually become explosive. Unlike Stello, he was not completely sure that Mattson was wrong. But he did tell reporters that the bubble would not explode that night. It would be several days at least before it would reach flammability. Questioned on how two different stories had come out of the NRC, Denton said, "There is no disagreement. I guess it is the way things get presented."

Those working at Three Mile Island did not stop to notice the panic of the press or of local citizens. Lead brick continued to arrive for stacking around the recombiner that it was hoped could take the hydrogen and oxygen of the bubble and make them into water. But that process could only start after the bubble was removed from the pressure vessel into the concrete containment.

Concern about explosions in the pressure vessel, which was stainless steel and eight inches thick, was minimal at Met Ed. In contrast with Bethesda, what worried Gary Miller and his operators was that there could be another blast in the concrete containment dome like the one that had produced the twenty-eight-pound pressure spike on Wednesday. The containment was designed and tested to withstand a pressure of only 60 pounds per square inch, and there was further worry about whether the concrete and steel dome had been weakened by the very high radiation inside. There was always the remote possibility that enough hydrogen—one expert had estimated that the potential blast force was equivalent to three tons of TNT—could breach the containment, releasing radiation to the environment. And this worry didn't hinge on oxygen's being evolved in the reactor, as Mattson's concern did. Whether or not there was oxygen in the core, there was plenty of it in the containment.

But all these concerns might be academic, since there was another way being tried to safely rid the reactor of its hydrogen bubble. Miller was attempting to use the primary cooling system to dissolve the hydrogen bubble slowly, carrying it out for venting through the pressurizer at a steady and safe rate. And it seemed to be working. Met Ed had not lied when Herbein said the bubble was shrinking.

And while important decisions were made about bubbles and recombiners and the presence of oxygen, in the Trailer City set up around the observation center, workers were coming on and off shift, eating in a mess tent where breakfast and dinner were both served constantly. Helicopters buzzed around the floodlit site as a radiation-monitoring plane took thousands of pictures of the area from high altitude. Nobody slept.

In Bethesda, the workers were eating too. Bernie Weiss had finally found a sub shop that delivered.

Roger Mattson and his team of experts were becoming increasingly agitated about the state of the bubble. Five percent oxygen was the amount needed to make a combustible mixture, and Tedesco's calculations showed that percentage approaching very quickly. Something had to be done immediately.

Bethesda didn't know what to suggest. Dozens of experts in every field were calling in, speculating about how to solve the problem. Much of the thinking centered around some chemical that might be injected into the cooling water to dissipate the bubble.

Not all of the people offering cures were on the list of normal NRC consultants. A few were engineers and chemists who had a suggestion and knew how to get the IRC number from a friend or acquaintance. And several of the callers were laymen who had learned one of the private NRC numbers from a radio call-in show broadcast from New York. "Listen, I know this may sound a little crazy," one of them suggested, "but I figure you've got to get rid of that bubble *and* get the thing shut down cold. They make a lot of dry ice in New Jersey and I bet you could get tons of the stuff and pack it around that dome to cool things down right away."

"The way I understand it," another said, "that thing's gonna blow up and all the radioactive particles are gonna get out. What you have to do is keep them in and don't let them get out. How about a big balloon around the whole building to catch the stuff and not let it out?"

"Yes, ma'am," replied a patient Bernie Weiss. "I'll give your suggestion to the men who are working on that problem."

Mayor Bob Reid, his evacuation plan finished, was running off leaflets on a mimeograph machine in the Middletown Borough Hall. There was commotion and panic in his community, but there wasn't much that he could do about it. The plan was ready. It might even work. And the problem of looters taking advantage of those who'd left was handled earlier in the evening: Bob Reid had told his police to shoot to kill. "Oh, I didn't really mean it," he said later, "but the word got around real fast and we never did have any looting."

Kevin Molloy did not have the serenity of Mayor Reid, but then he had not lived for forty-six years in the same town. Saturday had been very frustrating to the young Civil Defense director. First, he had called Amtrak—whose trains run along the riverbank past Three Mile Island—and asked that they cancel or divert the trains. Amtrak flatly refused.

Then, Molloy was feeling the effects of what he called a communication problem, and he wanted a solution to that problem. He wasn't really sure what was happening out at the Island. At 10:45 P.M. a message came across the PEMA teletype that NBC had said the

215

bubble had burst or was growing and that mass evacuations might soon take place. Alarmed, Molloy tried to call Thornburgh. He was told the governor was too busy to talk.

State Senator Richard Gekas was with Molloy in the communication center at the time. Gekas tried to call Thornburgh also but got the same response. Then Gekas tried Lieutenant Governor Scranton and was told that he too was busy.

"At that point," Molloy recalled, "Senator Gekas advised the lieutenant governor's representative . . . that if they did not get in touch with us with a little bit more information, that we would be performing our own evacuation at—I think it was nine o'clock the next morning."

Gekas and Molloy were deadly serious. Pennsylvania laws concerning evacuations—which were vague and probably had not been read until Friday night by anyone, including the governor—said that only Thornburgh could order an evacuation. But Molloy knew the difference between legal responsibility and actual responsibility.

"It all comes down on our shoulders," he said. "It's not really on the governor's shoulders or anyone else's. We're the ones that are going to ultimately have to make a decision."

Scranton called back, but the lieutenant governor was not in full command of the situation at that point and it was obvious to Molloy. Scranton's pleas were ineffective. An evacuation would go ahead unless the information situation improved. Scranton said he would be in to talk with Molloy in the morning.

It had been a bad day too for Saul Kohler, editor of the Harrisburg *Patriot*. Standing in line at the bank for $25 in "walking around" money, he saw the man in front of him withdraw $2,500 in cash.

In the evening Kohler found his synagogue had closed, just as half the churchgoers in town would find their places of worship shut the next morning. And then, at 2 A.M., he was aroused from a sound sleep by a frantic overseas call from his cousin, a dentist in Paris.

"Saul!" the cousin cried. "You're alive!"

20

Sunday Morning

"I believe," said Reverend Stephen Sparks, "we are living in the last days."

Reverend Sparks's congregation at the Glad Tidings Assembly of God murmured agreement.

"I'm not saying that to scare anybody," Sparks continued. "I'm just saying what the Bible tells us."

It was Sunday morning in Middletown and most of those who had stayed near their homes were in those churches that remained open for services.

At St. Paul's United Methodist Church in Newberrytown, Reverend Harold E. Millard looked out at his depleted congregation. "I don't know whether you're foolhardy or brave," he said. "We should remember in our prayers that there are many frightened people here."

"I lift up mine eyes to the hills," sang out Reverend Richard Deardorff, pastor of the Goldsboro Church of God, the church nearest to Three Mile Island. "May help come from the Lord, who made heaven and earth. As we gather here this morning, it is with fear and trepidation in our hearts. For we do not understand what is happening . . . I think God has been very kind to let us go as far as we have with His world. But I think now he is saying 'Be careful.' "

Evelyn Weirich stood to offer a prayer. Next to her sat three women who were the only others in the congregation that day. She said, "Maybe you are trying to tell the people of Goldsboro something. Maybe it's not too late for them to come to the foot of the cross and ask forgiveness. You know more than the technicians on Three Mile Island. It's all in Your hand, O Lord."

And Cora Shuler wept.

Colonel Oran K. Henderson walked into Kevin Molloy's office in the basement of the county building in Harrisburg. Molloy's army cot, his bed for the past four nights, was folded in the corner. Henderson had come in response to Molloy's urgent plea for information. An hour or so later, Young Bill Scranton arrived to talk with Molloy. There would not be an evacuation. There could not be an evacuation —not with the President coming to Harrisburg that very day. Scranton promised action on the information problem as a concession to the young Civil Defense director.

Word had come late the night before that President Carter would tour Three Mile Island on Sunday. Paul Critchlow was up most of the night getting ready for the visit. Critchlow and Thornburgh wanted the President to come. It would lend an aura of stability to the situation.

Charles Kline, chief of the Air National Guard fire department, voiced the feelings of many residents on learning that Carter was coming: "If they think it's safe enough for the President of the United States to come up, it's not too bad." That was just what Paul Critchlow wanted to hear.

The press in Harrisburg scrambled to prepare for Carter's visit. Here was a story they were used to covering. It was visual and not open to varied technical interpretations. But even the television networks had only a few hours to ready their coverage, so there was little time for the press to worry about personal danger, even after their scare of the night before.

NBC's Peter Hackes paid a local woman $20 for the use of her telephone to call in his radio story from Middletown. When he arrived to check out the line, he found another reporter already on the phone. "She probably got twenty bucks off him too," Hackes said.

Sunday Morning

The Boston *Globe* had withdrawn its lead science writer on Saturday night, telling Curtis Wilkie that they'd be using strictly wire-service coverage from then on. Wilkie, who had been a White House correspondent for the *Globe,* said, "Well, Carter's coming and I think I could handle that; I *have* done that from time to time." But Boston still said no. Wilkie was furious.

Mayor Bob Reid, dressed in his son's football jacket, spent the morning delivering evacuation-instruction leaflets to the people of his town. His constituents were scared and Reid wanted to reassure them. He took great pains to make sure that they understood what the leaflet was all about.

"Now, I want you to read this and tell me what it says, dear," Reid said to one elderly woman.

"Oh, Bobby," she said, "you know I can read."

"I know you can," Reid responded. "And I want you to tell me what it says right here. Read it to me."

The woman read, " 'In the event of an evacuation . . .' "

"Now, hold it right there," the mayor interrupted. "What does that mean? In the *event* of an evacuation. Does that mean we have an evacuation right now? No, it doesn't. That just means that we're ready, *in* the event. So you just sit here and be calm and don't you worry."

The engineers of Met Ed were sure the bubble was getting smaller. In fact, they had secret hopes that it would be gone completely in another day. And, like Vic Stello, they had no worries about a hydrogen explosion, since they felt there was no oxygen to speak of in the bubble.

Jim Panyard of the Philadelphia *Bulletin* spent part of the morning making a last try for information from Met Ed. Since Saturday morning, the utility had not been answering any questions from the press. But this time Panyard got the information he was seeking.

"I was told, 'Yes, the bubble is shrinking,' " Panyard recalled. " 'It's gone from 880 cubic feet to 560 to 420.' Later, when I called back to confirm the information, they said, 'I'm sorry, we can't talk

219

to you about the bubble. We are only giving out information on insurance. We aren't allowed to give any other information.' I think Met Ed was tickled with itself that the bubble was shrinking and decided to announce it."

But just as Met Ed was convinced that the bubble was no longer a threat, Roger Mattson's group calculated that oxygen had reached the critical 5-percent level. The bubble could detonate at any moment, they thought. All it needed was a spark.

Mattson left Bethesda for Harrisburg that morning, driving up with Chairman Hendrie to give Bethesda's version of the accident to Carter.

Jessica Tuchman Matthews started work before dawn, preparing a memo, with Jack Watson, explaining the situation in the reactor for Carter. Discussions were held with Brzezinski and Vice President Mondale. They were all concerned with how the visit would be viewed by the public and the press. Matthews was especially determined that the public not view Carter's visit as a sign that the accident was completely over—that there was still great reason for anxiety. She and Brzezinski urged the President to avoid being too optimistic in his public comments in Pennsylvania.

The presidential party arrived in Harrisburg at one P.M. With Carter and his wife Rosalynn were the sixteen reporters who always travel with the Chief Executive. A tour of the plant was scheduled, and all sixteen expected to go along. So did the three hundred reporters already in Pennsylvania. There was great animosity among the press when a single pool reporter was selected from among the sixteen. "The son of a bitch won't know a thing he's looking at," a local newsman groused.

Carter was no stranger to nuclear reactors. As a career naval officer, he'd served under Admiral Hyman Rickover, head of the Navy's nuclear program. In the helicopter going up to Pennsylvania, the President recounted his own meltdown experience, helping to dismantle a damaged reactor at Chalk River, Ontario, in 1950. "I'm not afraid of a little radiation," the President said. During the Canadian accident, he reportedly received 3 rems in an exposure period of only 90 seconds.

The press was aware of Carter's nuclear background too, though not in awe of it. During the 1976 election campaign, Congressman Harley Staggers of West Virginia tried to make a big point of Carter's scientific background when appearing with Carter in his home state.

"This Jimmy Carter's a smart man," Staggers said. "He went to Annapolis, studied how to be a nuclear fizzy-ist. I bet none a you out there is a nuclear fizzy-ist."

The press corps all put up their hands.

"Look at all the press with their hands up," Staggers crowed. "Y'all gonna go to hell for lyin'. Ain't none a you a nuclear fizzy-ist!"

On his arrival in Harrisburg, Carter met with Denton, Critchlow and others for a briefing in a large hangar where the radiological monitoring teams were stationed. Not all of the discussion was technical. "We stood around talking for a few minutes," Critchlow said, "and Carter said some incredibly insulting things about the governor, which we'll use when we run against him for President."

In the minutes before he was to brief the President, Denton heard from Stello and Mattson, seeking a consensus of NRC opinion. When Mattson gave Denton his estimate that the bubble could blow any time, Stello could not believe it. As Mattson recalled:

"Stello tells me I'm crazy, that he doesn't believe it, that he thinks we've made an error in the rate calculation . . . Stello says we're nuts and poor Harold is there, he's got to meet with the President in five minutes and tell it like it is . . . And here he is. His two experts are not together. One comes armed to the teeth with all these national laboratories and naval reactors people and highfalutin Ph.D.s around the country, saying this is what it is and this is his best summary. And his other expert, the Operating Reactor Division director, is saying, 'I don't believe it. I can't prove it yet, but I don't believe it. I think it's wrong.' "

Denton gave Carter the views of both his experts, trying to take a middle path himself. Carter listened and then advised keeping public statements conservative, no matter which man was correct.

"You've got to leave yourself some wiggle room," he said.

For the next two hours, Carter and Thornburgh toured the plant together, wearing yellow rubber shoe-wraps to keep from taking any radiation back off the Island with them. It would hardly do to have the President of the United States set off the frisker.

The pool reporter, who was the first newsman allowed in the control room of Unit 2 since the accident began, fully justified the fears of old hands covering TMI when he referred to Admiral Hyman Rickover as "Rick Over," with whom Carter had worked in the Navy some years back. "You remember good old Rick," chided one of the older science writers who had covered the Admiral since the launching of the *Nautilus,* the nation's first nuclear submarine, in 1954.

Back in Middletown, Carter and Thornburgh held a press conference together at the Borough Hall. True to his plan, Carter was conservative:

"I would like to say to the people who live around the Three Mile Island plant that if it does become necessary, your governor, Governor Thornburgh, will ask you and others in this area to take appropriate actions to ensure your safety. If he does, I want to urge that these instructions be carried out calmly and exactly as they have been in the past few days. This will not indicate that the danger is high. It will indicate that a change is being made in the operation of the cooling water system to permanently correct the present state of the reactor, and it is strictly a precautionary measure."

Mayor Reid, still dressed in his football jacket and sitting in the back of the hall, was approached by reporters a few minutes later. "Mr. Reid, when you saw the President of the United States standing right here in your Borough Hall, what did you think?"

"How short he was," Reid replied.

Later, the Middletown mayor had more to say about how Carter's visit affected the people of his town: "People weren't talking to one another. They were cooped up in their homes, and when he came, it seemed like everyone came out to see the President and it was really a shot in the arm . . . I saw humor but it was that type of humor

—Did you ever see people like they're ready to laugh but ready to cry at the same time?"

The tremendous worry in Bethesda that the hydrogen bubble might blow up was based entirely on the fact that Mattson's group of experts had been working on a false premise. Mattson was wrong and Vic Stello, alone, had been right. By 7 P.M., Stello was able to point out to Hendrie and Denton the flaw in Mattson's calculations. Mattson's group had based their work on an NRC regulatory guideline for hydrogen evolution in *unpressurized* containment structures. This bubble was in the pressure vessel at a thousand pounds per square inch. There was *no* oxygen.

"It was like a sudden light dawning," Denton said later. "By God, you are right, Vic . . . It had sort of been taken as a given that the oxygen generation rate for an open vessel applied here, and then everyone sort of went right from there . . . and the basic premise was incorrect."

At 8:30 P.M., Denton told Thornburgh that the bubble was down to 350 cubic feet. The next morning, it was less than half of that.

Met Ed went even further. George Troffer broke the utility's silence and claimed that the bubble was entirely gone. Later, under pressure from the White House, he tried to deny the story, but it was out. By Tuesday, Denton confirmed Troffer's claim.

Public concern continued to run high, even with the bubble no longer a threat, as smaller events continued to disturb the already-frightened people of central Pennsylvania: Rod Nordland, with his computer-controlled scanning radio, heard calls about another minor uncontrolled release at midweek; and the governor's recommendation that pregnant women and preschool children leave the area remained in effect while Thornburgh's staff looked desperately for a criterion for rescinding the recommendation.

Almost every day, Critchlow asked Denton if the advisory could be lifted, but there wasn't enough of an obvious improvement in the situation to justify such a move. Denton said they could rescind the advisory when the reactor was in cold shutdown, a state that would be reached in "a day or two." But days soon became more than a week and cold shutdown had not yet been reached.

On April 9, eleven days after asking women and children to leave the area around Three Mile Island, Thornburgh announced that he was lifting his "previous recommendations, advisories and directives." The pregnant women went back to their homes and the children back to their schools. There was still no cold shutdown.

21

Aftermath

Summer 1980. More than a year after the start of the accident at Three Mile Island. Trailer City was gone and the fields around the TMI observation center were returned to wheat and picnic tables, as they had been before the accident began. Most of the press was gone —on to another story in another place. And life in Goldsboro and Middletown had returned to relative normalcy.

But airline passengers, flying right over the nuclear plant on a long final approach into Harrisburg International Airport, could still see a wispy cloud of steam rising from one of the cooling towers of Unit 2. A wispy cloud of steam that was the only external evidence that the accident was still not over and that, a year after it all began, Unit 2 was still not in cold shutdown.

In the reactor 1,500 gallons of water had to be added each day to replace the 1,500 gallons that continued to leak daily into the containment sump. With the control rods gone—69 silver and boron rods melted by the intense heat of March 28, 1979—only the boron-laden cooling water kept the reactor from coming back to life. After a year, Unit 2 still relied on a single safety system in each area of plant operation. And these systems had been running for a year without maintenance in radiation fields far stronger than they were ever designed to withstand. If the last cooling pump or the last steam generator failed, then the accident at Three Mile Island would pick up right where it had left off in April 1979. Only, this time, there would be no safety systems at all to save the plant.

225

The decay heat that could cause a meltdown was much less than it had been a year before, and the hundreds of millions of curies of radiation inside the containment dome had dropped to only a few tens of millions of curies, but they were only kept from leaking through the concrete containment walls by a ventilating fan that had been running without pause or attention since the March before. The industry that had thought of accidents as "incidents" or "transients" that took only seconds or minutes to be resolved now had an accident that had lasted first days, then weeks, and then a whole year.

"Tell me when you're going to shut that damn thing off," Saul Kohler, editor of the Harrisburg *Patriot,* had asked Met Ed president Walter Creitz during the summer "after" TMI.

"We can't turn it off," Creitz replied.

Visitors to the observation center saw a film about the accident—Met Ed's version of the accident—that says, "the reactor is now in a state of cold shutdown." And through the ceiling-high windows of the observation center, they could see the wispy cloud of steam that proved this claim a lie. "We can't turn it off," Creitz had said.

There is no electricity generated at Three Mile Island. Unit 1, down for refueling at the start of the accident, but ready to go back into service ever since, is caught up in government red tape and local emotion, trapped in a thirteen-month process to decide when and if it can go back into service. And Unit 2 is out of action for years, or maybe forever. Cost of replacement power purchased by Met Ed from other utilities was $900,000 per day right after the accident began. And energy costs only go up. In replacement power, clean-up, rebuilding and legal expenses, the accident at Three Mile Island will probably cost Met Ed $3 billion. And without government or other subsidy, $3 billion is more than Met Ed can ever afford to pay, leaving TMI as a monument to a utility that was.

And subsidy of some type is on the way. After months of debate, the Pennsylvania Public Utilities Commission allowed General Public Utilities to pass on the costs of replacement power to the customers of its three subsidiaries—saddling citizens in New Jersey with part payment for mistakes made in another state.

But even with costs of replacement power covered and with

$300 million from the utility's insurers, Met Ed will still be left half a billion dollars short, money that may well come from every utility customer in the United States. In a triumph of free enterprise, *everyone* may pay for the mistakes of Three Mile Island.

In the year after the beginning of the accident at Three Mile Island, 1,700 technicians and engineers began the biggest clean-up operation in nuclear history, scrubbing buildings and equipment clean of radiation, literally with detergents and water, decontaminating 400,000 gallons of mildly radioactive water for eventual slow release into the Susquehanna and planning for the ultimate transportation of the reactor core and its highly radioactive waste material to its final storage place.

The year of scrubbing and clean-up was highlighted by delays, incompetence and poor planning, with workers becoming contaminated, stepping in puddles, working in unsafe conditions, spilling radioactive water and otherwise filling a regular spot on page two or three of most American newspapers. In all, ten workers had sustained high radiation doses, with one technician receiving eight times the maximum allowable exposure. At one point, some radioactive material was sent by mistake for burning in a local incinerator, only to be rejected at the dump because it was not properly packaged. In time, of course, this same crew of technicians and engineers will load up to 2,000 trucks with radioactive waste—and send those trucks off for a 2,700-mile trip on public highways to a nuclear waste disposal site in Washington State: 5 million miles of truck travel that will, it is hoped, be free of accident or danger to the public.

The actual physical danger posed by the accident at Three Mile Island, through its first year, has proved to be minor. Amazingly, it may have been made less severe by the very catastrophic nature of the damage to the reactor. The radioactive iodine that was of greatest concern to public health officials was never released in large quantities because it reacted with molten silver from what once had been the reactor's control rods. It now lies, decaying, in a slurry of silver iodide at the bottom of the reactor pressure vessel. What iodine survived the melting of the control rods was killed by sodium hydroxide sprays that went off accidentally after the hydrogen explosion that caused Wednesday's 28-pound pressure spike. If the accident had been not quite so bad—if the control rods had not melted or the containment sprays not been activated by the pressure spike —then the consequences to public health might have been much worse.

And what might have happened at Three Mile Island? This is the "what-if?" question that television loved so much. What if the very worst had happened? People would be dying. A Princeton University study estimated that releasing 10 percent of the iodine 131 would have resulted in 350 cases of thyroid cancer. This is cancer that could have been avoided by distribution of potassium iodide, not into homes, but at least to fire and police stations for ready access.

The potassium iodide, product of the FDA's frantic late-night calls, is still in a Pennsylvania warehouse, rapidly using up its shelf-life. Eventually, the FDA may reclaim the drug, having finally paid for it. But there are no drug companies waiting to produce potassium iodide pills. Mallinckrodt, the company that worked so hard to set up drug production in the middle of the night, has since sold its pharmaceutical division.

"My doctor assures me we will have a thousand cases of leukemia come out of this thing in twenty years," said *Patriot* managing editor Saul Kohler, "but he's an optimist. He thinks we'll have a cure by then."

What if the reactor had melted down? It almost did, with studies indicating that Unit 2 came within twenty minutes of a meltdown. But those same studies suggest that, in this specific case, such a meltdown wouldn't have been any more of a disaster than what did take place. Rather than leaving the reactor core for dangerous removal and transport across the country, it would have left it encased in a ball of volcanic glass a hundred feet below ground. Some experts think there could be no better way to store a poisoned reactor core.

And the experts feel that a meltdown would not have released significant amounts of radiation to the environment, but all do not agree on this point. Speculation about what if the fuel cycle had been further along, making more long-lived isotopes; what if the contain-

228

ment had been breached; what if the NRC had taken over the reactor, may never be answered. It *is* certain, though, that if in that moment at thirty-seven seconds after 4 A.M. on March 28, 1979, Faust, Fredrick, Scheimann and Zewe had responded to the lights and alarms by eating their packed lunches or taking a nap or doing almost anything else, there would have been far less damage to Unit 2. There might have been no accident at all.

The impact of TMI on government, industry and society has been varied, with the only consistent trend being that all sides are discouraged about the future. The rise of public distrust of nuclear power just following the accident has been tempered by events in Iran, the economy and by the steady increase in oil prices. While Gallup polls found that 66 percent of the population thought nuclear power to be unsafe in April 1979, only 50 percent continued to feel that way in January 1980. "I'm very pessimistic about the future," said Richard Pollock of the antinuclear group Critical Mass. "There has been very little movement. It is almost as if Three Mile Island never occurred." But Pollock's boss, Ralph Nader, has noticed another trend. "When I first began speaking against nuclear power," he said, "the audiences would ask me how I could prove that nuclear reactors were badly designed and poorly run. Now the audiences accept these problems without question and ask what alternatives we have to nuclear power."

If there seems to be little progress against nuclear power on our national scene—and a Swedish referendum in support of nuclear power in that nation and clear plans for nuclear expansion in France show support for nuclear power abroad—there is still a great deal of emotion against nuclear power in central Pennsylvania.

The citizens of Middletown and Goldsboro, people whose main concern with TMI before the accident was whether Met Ed would build them a recreation area, have banded together against reopening the plant. More than twenty antinuclear groups have been formed since the accident, and several of them promise civil disobedience if all else fails. "We no longer trust, we no longer believe," Reverend William Vastine of New Cumberland told NRC commissioners at a

229

meeting with local leaders. "If there's violence in the area . . . it's on your hands."

"You couldn't get ten people to a licensing hearing before the accident," TMI Alert head Chris Sayre claimed, "but now they get five hundred local people for any hearing you want."

Local distress centers around the long-term effects of living near TMI. "How can I raise children here, in sight of that place, if it is going to make my grandchildren or great-grandchildren into freaks?" asked one local mother. The birth of three deformed children in the area in early 1980 caused a tremendous public outcry against the plant and against nuclear power in Pennsylvania. Statistically the birth defects were not significant—were very unlikely to have been caused by TMI—but they showed the level of feeling in the community. "The emotional trauma is ten times as important as the physical effects," said Dr. Joseph Leaser, a local physician, "and the people are not addressing that problem. It's like a powder keg waiting to explode."

Beyond the fear of the mothers of Middletown and Goldsboro, another emotion has run very strong in central Pennsylvania in the year since TMI—greed. In the wake of the accident, fifty-six class-action suits were filed against Met Ed, General Public Utilities, Babcock & Wilcox and others. The claims total in the billions of dollars. Thousands of depositions have been taken in the suits, leaving nobody in the area surrounding Three Mile Island without an opinion or hope of monetary gain.

The nuclear industry, too, is depressed. Poor economics, bad publicity and fears of increasing regulation in the wake of Three Mile Island resulted in the cancellation of orders for eleven nuclear power plants in 1979. No new plants were ordered that year. Work on plants already under construction was slowed and plans were even made to change some proposed nuclear plants to coal-fired ones. Nearly $2 million spent on pronuclear publicity by the Atomic Industrial Forum has been more than offset by the clean-up problems of TMI, smaller accidents at other nuclear facilities and a federal indictment charging Commonwealth Edison Co. of Chicago with lying to the NRC about security precautions at one of its nuclear plants.

Still, some good has resulted from the accident. The nuclear industry has begun to police itself better, setting up two industry bodies charged specifically with researching and monitoring safety in nuclear plants.

As a follow-up to Three Mile Island, both Met Ed and the NRC have chosen to place most of the blame for the accident on the operators—in direct opposition to their early stand. But operators are cheap to hire and train, so it is easier to blame the men than the machine. Even the machine has been changed some, though. Soon every reactor in the country will have a gauge to tell how full the cooling system is, not just how full the pressurizer is. And every Electromatic valve will have an instrument to show whether it is open or closed. Despite such instrumentation, though, failure of the computer on a Babcock & Wilcox reactor at Crystal River, Florida, has since resulted in failure to close an Electromatic valve, with the loss of 43,000 gallons of cooling water, though without other damage.

The NRC has come under fire for its performance at TMI. First a presidential commission and then a commission set up by the NRC itself have suggested total reorganization of the regulatory agency that had thought its job to be more one of writing regulations than regulating. Such a reorganization, if it comes, is still in the future. In the meantime the NRC muddles on with contradictory ideas, requirements and plans for increasing safety. The agency has produced a 200-point "action plan" that will cost $30 million per power plant to implement, yet has been found to be unsuitable and disorganized by both the nuclear industry and antinuclear groups—a rare agreement of these two groups. The NRC can't even agree on where its best brains should be stationed to react to the next emergency to avoid the factionalization of opinion and information evident at TMI. Vic Stello supports sending the NRC to the reactor, while Roger Mattson gives unqualified support for setting up headquarters in Bethesda.

*   *   *

Everyone involved in the accident at Three Mile Island has been changed somewhat by the experience. Walter Creitz is no longer president of Metropolitan Edison. Lee Gossick is no longer with the NRC. Joseph Hendrie has been demoted from Chairman of the NRC to being just one of its commissioners. Dr. Gordon MacLeod is no longer Pennsylvania's Secretary of Health. And a few dozen reporters who never thought anything about nuclear power before Three Mile Island now find themselves covering the story full-time.

Nuclear power is a popular subject for the press, and every leaking let-down line or careless worker is the subject of a newspaper or television account. But the press hasn't learned as much as it thinks it has about covering nuclear accidents. In October 1979 a steam leak at a reactor in Minnesota brought a hundred TMI veterans together again. The Associated Press, in its account of the accident, noted that a general emergency was declared at 2 P.M., but that there was no measurable radioactivity at 4:30 P.M. Just as at Three Mile Island there was no mention that a general emergency indicated a dangerous release of radiation and no follow-up on the fact that, while there was no radiation at 4:30 P.M., there may have been a lot of radiation at 2 P.M.

A final look at the performance of government during the accident at Three Mile Island shows the same fear of honesty—if honesty carries any sign of weakness with it—that had been characteristic of Met Ed in the accident's first two days. Would it have been fatal for the White House, the state house or the NRC to admit that they did not know everything or to admit that mistakes had been made? What if the utility and the government had told the truth all the time? Would that have caused panic? The White House told its commission studying the accident to see "how well the public's right to know was served during the accident." The major effort to restrict public information was made *by* the White House. The governor's office set up a rumor-control line that made no attempt to control rumors. On April 4 Met Ed had a meeting to plan how best to avoid the press in future, even to the extent of deciding when it was advisable to lie.

Looking back at the accident and at the state he had covered for years in Harrisburg, Jim Panyard of the Philadelphia *Bulletin* said, "The NRC and the Civil Defense were about as prepared to cope with this as you would be to keep your dog from running in the street and getting hit by a car."

Have things really improved?

A month after TMI, an electrified Amtrak train was speeding from New York to Harrisburg. Conversation among some passengers going home turned to what it had been like to live in sight of the cooling towers for those harrowing days. Sweeping west, up the green Susquehanna valley with its old farms and fertile fields, the train turned in, parallel to the river, running up the last stretch into the state capital. Rolling between Route 441 and the river, the train passed between the observation center and Three Mile Island.

A large black woman sat on the left side of the coach, looking up at the 300-foot cooling towers and the 100-foot concrete containment domes. Her two-year-old daughter lay sprawled on the seat next to her, asleep.

"Is that the place—that Three Mile Island?" she asked the men across the aisle, who had been talking of their fears on the way down.

It was.

Quickly, the mother took off her sweater and placed it over the sleeping child.

233

Glossary

Auxiliary building. A structure housing a variety of equipment and large tanks necessary for the operation of the reactor. These include makeup pumps, the makeup and waste-gas decay tanks, and the reactor coolant hold-up tanks.

Babcock & Wilcox (B&W). The company that designed and supplied the TMI Unit 2 reactor and nuclear steam supply system.

Background radiation. Radiation arising from natural radioactive materials always present in the environment, including solar and cosmic radiation and radioactive elements in the upper atmosphere, the ground, building materials and the human body.

Building shine. Gamma radiation—not visible to the naked eye—strong enough to pass through the concrete containment dome and endanger plant workers.

Bureau of Radiological Protection (BRP). A division of Pennsylvania's Department of Environmental Resources. BRP is the state's lead agency in monitoring radiation releases from nuclear plants and advising the Pennsylvania Emergency Management Agency during radiological emergencies.

"Burned out." A slang term used by nuclear workers for overexposure to radiation, exceeding 3 rems exposure per quarter-year, the allowable level set by the Environmental Protection Agency.

Burns & Roe. Architectural and engineering firm responsible for the design of TMI Unit 2.

Candy cane. The section of pipe carrying water from the reactor to a steam generator.

Cavitation. A malfunction of the main coolant pumps in which steam voids, or cavities, are pushed through the cooling system.

Chain reaction. A self-sustaining reaction; occurs in nuclear fission when the number of neutrons released equals or exceeds the number of neutrons absorbed plus the neutrons which escape from the reactor.

234

Glossary

Cladding. In a nuclear reactor, the metal shell of the fuel rod in which uranium oxide pellets are stacked.

Condensate polisher. A device that removes dissolved minerals from the water of the feedwater system.

Condensate pumps. Three pumps in the feedwater system that pump water from the condensers to the condensate polishers.

Condenser. Device that cools steam to water after the steam has passed through the turbine.

Containment building or dome. The structure housing the nuclear reactor; intended to contain radioactive solids, gases and water that might be released from the reactor vessel in an accident.

Control rod. A rod containing material that absorbs neutrons; used to control or halt nuclear fission in a reactor.

Core. The central part of a nuclear reactor that contains the fuel and produces the heat.

Core flood tank. A 500,000-gallon tank of borated water used to cool the reactor core in a major accident.

Critical. Term used to describe a nuclear reactor that is sustaining a chain reaction.

Curie. A unit of the intensity of radioactivity in a material. A curie is equal to 37 billion disintegrations each second.

Decay heat. Heat produced by the decay of radioactive particles; in a nuclear reactor this heat, resulting from materials left from the fission process, must be removed after reactor shutdown to prevent the core from overheating. See *radioactive decay.*

Design basis accident (DBA). Hypothetical accident evaluated during the safety review of nuclear power reactors. Plants are required to have safeguards that will ensure that radiation releases off-site will be within NRC limits should any of these accidents occur.

Dosimeter. A device used to measure personal exposure to radiation.

Electromatic valve. A trademark of Dresser Industries, Inc. Also known as a Pilot Operated Relief Valve (PORV) designed to automatically relieve excess pressure in the primary-cooling system at 2,155 pounds per square inch.

Emergency core cooling system (ECCS). A backup system designed to supply cooling water to the reactor core in a loss-of-coolant accident.

Emergency feedwater pumps. Backup pumps intended to supply feedwater to the steam generators should the feedwater system fail to supply water. Also called auxiliary feedwater pumps.

Failed fuel. Structurally damaged fuel elements whose ruptured cases allow uranium dioxide to come into contact with primary-coolant water.

Feedwater pumps. Two large pumps capable of supplying TMI-2's two steam generators with up to 15,500 gallons of water a minute.

GLOSSARY

Feedwater system. Water supply to the steam generators in a pressurized water reactor that is converted to steam to drive turbines; part of the secondary loop.

Fission. The splitting apart of a heavy atomic nucleus, into two or more parts, when a neutron strikes the nucleus. The splitting releases a large amount of energy.

Fission products. Radioactive nuclei and elements formed by the fission of heavy elements.

Fuel damage. The failure of fuel rods and the release of the radioactive fission products trapped inside them. Fuel damage can occur without a melting of the reactor's uranium.

Fuel melt. The melting of some of the uranium oxide fuel inside a reactor.

Fuel rod. A tube containing fuel for a nuclear reactor.

Gamma rays. High-energy electromagnetic radiation, stronger than X-rays, that penetrate very deeply into body tissues.

General emergency. Declared by the utility when an incident at a nuclear power plant poses a potentially serious threat of radiation releases that could affect the general public.

General Public Utilities Corporation (GPU). A utility holding company; parent corporation of the three companies that own TMI.

Genetic defects. Health defects inherited by a child from the mother and/or father.

Half-life. The time required for half of a given radioactive substance to decay.

Health physics. The practice of protecting humans and their environment from the possible hazards of radiation.

High-pressure injection (HPI). A pump system, capable of pumping up to about 1,000 gallons of cooling water a minute into the reactor coolant system; part of the emergency core-cooling system.

"Hot lab." The laboratory on TMI used for measuring radioactivity of coolant water.

Iodine-131. A radioactive form of iodine, with a half-life of 8.1 days, that can be absorbed by the human thyroid if inhaled or ingested and cause noncancerous or cancerous growths.

Isolation. Condition intended to contain radioactive materials released in a nuclear accident inside the containment building.

Krypton-85. A radioactive noble gas, with a half-life of 10.7 years, that is not absorbed by body tissues and is soon eliminated by the body if inhaled or ingested.

Let-down line. A means of removing water from the reactor coolant system.

Loss-of-coolant accident (LOCA). An accident involving a broken pipe, stuck-open valve, or other leak in the reactor coolant system that results in a loss of the water cooling the reactor core.

Makeup system. A means of adding water to the reactor coolant system during normal operation.

Makeup tank. A storage tank in the auxiliary building which provides water for the makeup pumps.

236

Meltdown. The melting of fuel in a nuclear reactor after the loss of coolant water. If a significant portion of the fuel should melt, the molten fuel could melt through the reactor vessel and release large quantities of radioactive materials into the containment building.

Metropolitan Edison Company (Met Ed). Operator and part owner of the Three Mile Island nuclear power plant.

Millirem. One-thousandth of a rem; see *rem.*

Natural convection (or *natural circulation*). The circulation of water without pumping by heating water in the core and cooling it in the steam generator.

Neutron. An uncharged particle found in the nucleus of every atom heavier than ordinary hydrogen; neutrons sustain the fission chain reaction in nuclear reactors.

Noble gases. Inert gases that do not react chemically and are not absorbed by body tissues, although they may enter the blood if inhaled. These gases include helium, neon, krypton, xenon and radon.

Nuclear Regulatory Commission (NRC). U.S. agency responsible for licensing and regulation of nuclear reactors. It took over many functions of the Atomic Energy Commission, which was dissolved in 1975.

Pennsylvania Emergency Management Agency (PEMA). Agency responsible for the state's response to natural and human-made disasters.

Plume or Gaussian plume. A cloud of radioactive gas and particles spreading downwind from a nuclear accident.

"Poisons." Materials that strongly absorb neutrons; used to control or stop the fission reaction in a nuclear reactor.

Potassium iodide. A chemical that readily enters the thyroid gland when ingested. If taken in a sufficient quantity prior to exposure to radioactive iodine, it can prevent the thyroid from absorbing any of the potentially harmful radioactive iodine-131.

Pressure vessel. The steel tank containing the reactor core; also called the reactor vessel.

Pressurized water reactor. A nuclear reactor system in which reactor coolant water is kept under high pressure to keep it from boiling into steam.

Pressurizer. A tank that maintains the proper reactor coolant pressure in a pressurized water reactor.

Primary system. See *reactor coolant system.*

Radioactive decay. The spontaneous process by which an unstable radioactive nucleus releases energy or particles to become stable.

Radioactivity. The spontaneous decay of an unstable atom. During the decay process, ionizing radiation is usually given off.

Radiolysis. The breaking apart of a molecule by radiation, such as the splitting of water into hydrogen and oxygen.

Rad-waste tank. A storage place for spent reactor fuel.

GLOSSARY

Reactor (nuclear). A device in which a fission chain reaction can be initiated, maintained and controlled.

Reactor coolant pump. One of four large pumps used to circulate the water cooling the core of the TMI-2 reactor.

Reactor coolant system. Water that cools the reactor core and carries away heat. Also called the primary loop.

Reactor vessel. See *pressure vessel.*

Rem. A standard unit of radiation dose. Frequently radiation dose is measured in millirems for low-level radiation; 1,000 millirems equal one rem.

Respirator. A breathing mask that filters the air to protect against the inhalation of radioactive materials.

Saturation temperature. The temperature at which water at a given pressure will boil; the saturation point of water at sea-level is 212 degrees F.

Scram. The rapid shutdown of a nuclear reactor, by dropping control rods into the core to halt fission.

Secondary system. See *feedwater system.*

Site emergency. Declared by the utility when an incident at a nuclear power plant threatens the uncontrolled release of radioactivity into the immediate area of the plant.

Solid system. A condition in which the entire reactor coolant system, including the pressurizer, is filled with water.

Steam generator. A heat exchanger in which reactor coolant water flowing through tubes heats the feedwater to produce steam.

Steam table. A chart used to determine the temperature at which water will boil at a given pressure.

TMI. Three Mile Island, site of two nuclear power reactors operated by Metropolitan Edison Company.

Transient. An abnormal condition or event in a nuclear power system.

Trip. A sudden shutdown of a piece of machinery.

Turbine building. A structure housing the steam turbine, generator and much of the feedwater system.

Uranium dioxide. A chemical compound containing uranium and oxygen that is used as a fuel in nuclear reactors.

Venting. The intentional release of gaseous radioactive material into the environment.

Waste-gas decay tank. One of two auxiliary building tanks in which radioactive gases removed from the reactor coolant are stored.

Xenon-133. A radioactive noble gas, with a half-life of 5.3 days, that is not absorbed by body tissues and is soon eliminated by the body if inhaled or ingested.

Zircaloy-4. A zirconium alloy from which fuel-rod cladding is made.

Index

239

INDEX

INDEX

245

About the Author

MARK STEPHENS was born in Ohio. He has worked as a broadcast newsman and as a free-lance journalist both in the United States and abroad. His byline has appeared in newspapers throughout the United States and including the Boston *Globe* and the Los Angeles *Times*.

He has worked as a communication consultant to both government and industry and, most recently, served as a public information specialist for the President's Commission on the Accident at Three Mile Island.

He received a Ph.D. in public affairs communication at Stanford University, where he teaches journalism. Stephens now lives in Palo Alto, California.